The Biology of
Estuarine
Management

The Biology of Estuarine Management

JAMES G. WILSON

London   Sydney   New York
CROOM HELM

First published in 1988 by
Croom Helm Ltd
11 New Fetter Lane, London EC4P 4EE

Croom Helm Australia
44–50 Waterloo Road
North Ryde, 2113, New South Wales

Published in the USA by
Croom Helm in association with
Chapman and Hall
29 West 35th Street, New York, NY 10001

© 1988 James G. Wilson

Printed in Great Britain by
St Edmundsbury Press Ltd
Bury St Edmunds, Suffolk

ISBN 0 7099 5208 2

British Library Cataloguing in Publication Data

Wilson, James G., 1951–
    The biology of estuarine management.
    1. Estuary ecosystems.   I. Title
    574.5'26365

    ISBN 0–7099–5208–2

Library of Congress Cataloging in Publication Data

Wilson, J. G. (James Gow)
    The biology of estuarine management/
    James G. Wilson.   p.   cm.
    Bibliography: p.
    Includes index.
    ISBN 0–7099–5208–2
    1. Estuarine ecology.   2. Estuarine area
    conservation.   I. Title.
    QH541.5.E8W55   1988   574.5'26365—dc19

CONTENTS

PREFACE

Estuaries are the interface between man and the sea, and they are the channels for the impact of man on the marine environment. Because they are to a greater or lesser extent connected to the sea, they have traditionally been regarded as part of that seemingly infinite resource, or at best an open-ended means of access to it. This approach has led to conflicts between the users of the estuary, and with the increase not only in population but more particularly in the developed countries in manufacturing output, these conflicts have become more and more acute .

The estuary should always be regarded as a resource, and a finite resource at that, and the problem in management is to optimise the use of that resource. It is clearly wasteful and inefficient not to use it to its full capacity, and this includes both overuse of the system, such that the whole thing collapses, and underuse, in which there is still spare capacity for one use or another. The objective of this book is to explore the uses to which estuaries are put and the means by which the performance of the system under load may be assessed.

It seems appropriate here to mention that although this book will be talking about estuaries, the majority of it will be applicable also to lagoons, semi-enclosed bays and other such systems. In addition some topics, which do not in the majority of cases relate directly to estuaries, such as the disposal of sewage sludge or dredge spoil are also included as they are major management considerations.

As will be discussed more fully in the full

text, "estuary" is itself a term capable of being defined in a number of ways, most of which will not exclude those systems mentioned above. They are subject to the same sorts of conflicts and suffer from the same sorts of problems as estuaries and so the management of them can be approached in the same sort of way.

This book is written from the biologist's viewpoint, but with the emphasis on systems and processes rather than on the particular species themselves. Species have been considered in a functional way by what they do in the estuarine system rather than by their phyletic affinities. It is intended to be of interest and of use to all those interested or involved in estuaries. These include not only the professionals, such as the managers, engineers and scientists (including biologists), but also students and the public. Environmental awareness has greatly increased in the past few years and the so-called "Green" movement is a political force to be reckoned with in many countries. People no longer accept that there is no alternative to pollution and this attitude is reinforced by the comparatively short history of the decline of environmental quality, most of which has occurred within living memory. That final statement is difficult to prove, usually because hard scientific evidence for the period in question is lacking, yet as long as the public perception is that this is so, there will be calls for something to be done about it. To what extent it is feasible or even possible for something to be done is a judgement that the reader should hopefully be better able to make by the end of this volume.

Why should we bother with conservation and protection of the natural environment at all? There are two reasons, one philosophical and one practical. The first reason is that nature itself has an intrinsic value, that individual species are worth preserving for their own sake. Whether one invokes a supernatural power as justification for this or not is hardly relevant, although it is interesting to note the shift in the last twenty years or so away from the argument expressed in Genesis 1.28, that man should "subdue the earth and have dominion over all living things", to that in Genesis 2.15 in which man's role is more that of a steward whose job it is to "dress and keep". The second argument is the practical one that these things do have a value to man, either directly at the present in the form of a commercial crop for example, or that they may at some time in the future be required. The common reasoning for the latter argument is the necessity to maintain the gene pool, on the grounds that such diversity has evolved for a purpose, such as resistance to disease, and this may be useful or even essential at some future date.

Finally it should not be forgetten that man is an integral part of the estuarine system; any damage to, or alteration of that system will ultimately affect man as well. The point of using indicators is that they can serve as an early-warning so that is is possible to do something about it before the effects are felt by man. It is in our own interests, in the broadest sense, to get it right.

Acknowledgements

May I thank Dr. B. Magennis of Trinity College, Dublin for permission to use some of her data (Figure 4.6), and Dr. P.C. Nicholas of the Welsh Water Authority and his publishers, Plenum Press, for permission to use Figure 5.1. Other figures have been prepared by myself with the valuable assistance of my wife, Elizabeth Wilson, whose contribution to the volume, not least through her forbearance and understanding, was above and beyond the call of duty.

I should like to thank all my colleagues in the Environmental Sciences Unit, TCD and the Irish Estuarine Research Programme for their encouragment and suggestions. In particular, I should like to thank my colleague Dr. D.W. Jeffrey for his assistance and guidance in so many aspects of this work. Some acknowledgement is also due to all those estuarine workers, of whatever persuasion, whose comments and observations led to the idea for this volume, and I would greatly appreciate if any omissions or mistakes in the text, or comments on the subject matter could be pointed out by the reader.

LIST OF TABLES AND FIGURES

Tables

ABBREVIATIONS AND ACRONYMS

ADP	Adenosine di-phosphate
AEC	Adenylate energy charge
AHH	Aryl hydrocarbon hydroxylase
AMP	Adenosine mono-phosphate
ATP	Adenosine tri-phosphate
BAT	Best available technology
BCF	Bioconcentration factor
BOD	Biological oxygen demand
BMPA	Best practical means available
BPI	Benthic pollution index
BQI	Biological quality index
BT	Burrowing temperature
CNS	Central nervous system
COD	Chemical oxygen demand
DDT	Dichloro-diphenyl-trichloro-ethane (1,1,1-Trichloro-2,2-bis(p-chlorophenyl)ethane) (also DDD, DDE)
DO	Dissolved oxygen
DOC	Dissolved organic carbon
DOM	Dissolved organic matter
EEC	European Economic Community
EIA	Environmental impact assessment
EIS	Environmental impact statement
EPA	Environmental Protection Agency
EQO	Environmental quality objective
EQS	Environmental quality standard
FT100	Financial Times 100 (share index)
GESAMP	(Joint) Group of experts on the scientific aspects of marine pollution
HCH	hexachlorocyclohexane
IMCO	Internationl Maritime Consultative Organisation
LC	Lethal concentration
LD	Lethal dose
LDC	London Dumping Convention
LOT	Load on top
MAC	Maximum allowable concentration
MARPOL	Oslo and Paris Commissions (marine

	pollution)
MATC	Maximum allowable toxicant concentration
MBAS	Surface active substances
NGE	Net growth efficiency
NWC	National Water Council
O:N	Oxygen:nitrogen ratio
OTEC	Ocean thermal energy conversion
P:B	Production:biomass ratio
PCB	Polychlorinated biphenyl
PCP	Pentachlorophenol
PHC	Petroleum hydrocarbon
PLI	Pollution load index
PNA	Polynuclear aromatic
POC	Particulate organic carbon
POM	Particulate organic matter
RA	Respiration assimilation
RORO	Roll-on, roll-off
RPD	Redox potential discontinuity
RSA	Rank species abundance
SAB	Species:abundance:biomass
SCE	Sister chromatid exchange
TBT	Tributyl tin
TOC	Total organic carbon
UES	Uniform emission standards
UNESCO	United Nations Educational, Scientific and Cultural Organisation
UV	Ultra-violet
WSF	Water-soluble fraction

INTRODUCTION

IMPORTANCE OF ESTUARIES

The term "estuary" itself is derived from the latina aestus, meaning tide and hence aestuarium, a tidal arm of the sea. However, the term as presently employed adds to this the concept of freshwater input, to create a region of brackish water. Remane (1971) recognised seven types of brackish-water environment, namely (i) brackish seas such as the Baltic, (ii) estuaries where a river meets the sea, (iii) fjords in mountain valleys, (iv) lagoons, where the free access to the sea has been cut off by a sand bar, (v) shore pools, both on rocky and on sandy or muddy shores, (vi) salt marshes, at high level but still within the range of the tides, and finally (vii) coastal ground waters, where freshwater input comes from groundwater seeping out into the sea.

Within the last thirty years or so, there have been attempts to define the term with a little more scientific rigour according to physical, chemical or even biological characteristics (see Chapter Two), but the overall consensus differs little in principle from the rather vague notion above. Hedgepeth (1983) in his consideration of the subject, lumped together Remane's (1971) three major types of brackish-water, namely the brackish seas, estuaries and lagoons, largely on functional grounds, while at the same time emphasising the scope of studies that have at one time or another come within the ambit of the estuarine researcher. While each estuary is unique with its own set of characteristics, there are nevertheless some general statements that can be made about them as a whole.

1

Introduction

Early biological investigations into estuaries tended to focus on the problems of life in estuaries, notably the aspect of osmoregulation to cope with the salinity changes caused by the mixing of fresh- and salt-water, but more recently attention has been focused on a whole range of estuarine processes, particularly those of importance to man's activities in and around estuaries (e.g. McLusky, 1981; Kennish, 1986; Knox, 1986, in the Bibliography at the end of this Introduction).

Man uses the estuary as a habitat, as a source of food, for transport, as a means of disposing of wastes and more recently for recreation and enjoyment. These activities have threatened and in many cases altered the characteristics, physical, chemical and biological, of the estuary to the extent that the very use to which it is being put is threatened as well as creating conflicts between incompatible activities. The estuary is a limited resource and the aim of management should be to optimise the exploitation of this resource.

As the pressure on the resource has increased, so have the difficulties of management increased. There have been considerable advances in our ability to detect pollution stress and its effects at different levels of the system (e.g. Bayne et al., 1985) and in many countries public awareness of and interest in pollution has reached the stage where it is of considerable political importance. Along with this has been a broadening of the pollution concept. Where once it could be thought of solely in terms of the oxygenation status of the water, the emphasis in many cases has shifted to the persistent, accumulable, long-term compounds and elements and to sub-lethal carcinogenic or teratogenic effects.

The effect of this has been to focus more attention on the biological system, its role in the estuary, how it is affected by pollution, and the consequences of any effect on the important part the estuary plays in relation to man's activities and to the adjoining systems.

DYNAMIC ROLE

The importance of the estuary to many of man's activities is due in no small part to its highly dynamic nature.

Its hydrological regime with the combination of rapid salinity fluctuations, high water energy and

2

the pattern of sediment suspension and deposition, are major factors in the geochemical fluxes of many elements in the system. Through their action substances, not only contaminants such as heavy metals but also nutrients like nitrogen and phosphorus, are both removed from and renewed into the system. The sediment in particular tends to act as a sink for many materials, and through the decomposition of organic matter plays a major role in the energy flow through the system both directly and indirectly with the release of nutrients for primary production (Knox, 1986).

Estuaries are also highly dynamic biologically, and are areas of naturally high productivity. This is due in part to hydrodynamic factors, including freshwater input of nutrients but principally the to system's ability to trap and re-release nutrients. Transformation and speciation are largely biologically mediated: the rate at which nutrients are made available depends on the degree and type of microbial activity in the sediments, and in the case of salt-marshes, the plants actually act as pumps to draw the nutrients out of the substrate (Kennish, 1986). This high primary productivity can be passed up the system, with the result that estuaries not only give high yields of animals such as oysters and mussels, but also act as significant hosts to great numbers of juvenile fish on the one hand and birds on the other, both attracted by the wealth of food available.

As a consequence of all this activity, the estuary assumes an importance to the marine environment out of all proportion to its physical dimensions. Not only does it act as an important mediator of man's impact on the sea, it also acts as a focus for many of the coastal processes including geochemical cycling, energy flow and stock recruitment. Again, the estuary's role in this regard has only begun to be appreciated, but it has given added significance to the importance of management

SCOPE OF VOLUME

Management goals and practices have grown up and changed in response to the demands and perceived objectives of the time. In past years, the difficulties in dealing with the biological system, notably with regard to areas such as predictive modelling, have meant that the biological side was

Introduction

either dealt with empirically, or ignored altogether
(O'Kane, 1980). However, the shift toward biological
considerations in recent years means that such
attitudes are no longer possible, and the biological
system, its characteristics, processes and responses
have now to be taken into account.

This volume is intended as an introduction to
the part the biological system plays within the
whole, to the way it responds to the various
pressures imposed on it by the demands of the
various estuarine uses, and to how these responses
are being harnessed to quantify the degree of impact
on the estuary. In this, it follows the recent trend
towards the process-orientated approach, in which
the emphasis is on what an organism does within the
system rather than its place in the phyletic or
evolutionary scheme of things.

Chapter One

Chapter One describes how the various uses of an
estuary have evolved over the years. There are few
which have not been affected to a greater or lesser
extent by man's activities. Industrialised estuaries
show the greatest changes, in everything from
physical shape to biological quality status, but
even non-industrialised, non-settled estuaries in
undeveloped countries can be affected by actions
remote from the estuary itself - for example
forestry management practices within the river
catchment area affect river flow and sediment load
into the estuary.

The principal usages of the estuary are shipping
or transport, fishing, waste disposal both domestic
and industrial, and amenity value which includes
both the recreational and conservation aspects. Each
of these can make a substantial contribution to the
economy of the area, and the management objective
must be to achieve the optimum balance between
conflicting demands.

Chapter Two

In Chapter Two we take a look at the various
aspects of the estuarine system, using as a
framework the various definitions and classification
schemes that have been put forward. It was felt that
a comprehensive treatment of the physical and
chemical systems could not be attempted in a volume

of this size, and so these topics have been outlined rather than gone into in any detail. For these, the reader is referred to recent volumes such as those of O'Kane (1980), Bowden (1983), Head (1985) and Dyer (1986) and to others in the Bibliography at the end of this Introduction. The emphasis has rather been on the biological system, through the geochemical cycles especially of nitrogen, phosphorus and sulphur, the productivity and the other salient characteristics of the organisms which make up the estuarine biota. An understanding of these processes and characteristics is fundamental not only to an understanding of the system's response to man's activities but also to an understanding of the basis and the limitations of how the response is measured and quantified.

Chapter Three

Chapter Three goes through some of the different pressures that are put on estuarine systems, and likewise the emphasis here has been rather selective, with the greater part of the chapter devoted to waste disposal: the types, the characteristics and the effect on organisms. While some of the disposal options are discussed, wastewater treatment as such has again been largely dismissed, as this is a subject which has been dealt with at greater length and in much more depth than would be possible here (see e.g. Curds and Hawkes (1983) or Gray (1988) for a comprehensive outline).
It is intended in this chapter to reflect the balance of concern at the present time. A major concern in almost all estuaries is sewage disposal, in which the problem has traditionally been perceived as one of biological oxygen demand (BOD) and olfactory and visual nuisance. However, there is increasing awareness of the less immediately noticeable dangers, the pathogens and nutrients in sewage and the persistent, accumulable pollutants such as heavy metals and organochlorines, and carcinogens and genotoxins such as many of the modern synthetic organochemicals and radioactive isotopes in industrial discharges.
The way in which the system deals with the different problems posed by the different pollutants is also discussed along with the pathways by which they may be transferred through the system as well as the effects they have on the way.

Introduction

Chapter Four

In Chapter Four there is an account of the
consequences of pollution, and of the variety of
methods that have been employed to detect it. Gross
pollution is immediately obvious even to the
untrained observer, and no special methods are
necessary at this stage. Once such a state is
reached, remedial action is a difficult, costly and
lengthy process, and it is clearly in nobody's
interest to have the estuary acting as little more
than an open sewer. However, it would be equally, if
not more difficult to prevent any of man's actions
from impinging on the estuary, so clearly a balance
has to be struck between the two extremes.

Such a balance would permit the use of the
estuary up to that point at which damage or
deleterious effects - by which the term "pollution"
itself is defined - are caused, and it is this
balance that we want to be able to say has been
reached. Consequently, we have to be able to detect
subtle changes in the biological system which will
tell at what point the system is on the scale from
completely pristine to grossly polluted. This
implies a quantitative and objective measure of the
degree of contamination or system damage.

There are a whole range of such tests - indeed
one may be forgiven the observation that there
almost as many tests as there are scientists working
on the problem - ranging in scale from those that
consider the whole estuarine system to those which
measure response at sub-cellular level, and examples
are given of these, the conditions in which they
have been employed and some of the limitations to
which they are subject.

Chapter Five

This last chapter is devoted to management
considerations, again largely approached in the
light of the characteristics of the biological
system. It sets out the goals of management, the
ways in which they may be set up and monitored and
the considerations, economic and practical, that
have to be taken account of.

The central theme is that of the quality
standards: how they are derived, the different
standards required depending on the management
objective, and an outline of the major international

legislation under which authorities are increasingly
being required to operate.
    Finally an attempt is made to look into the
future of estuaries and the changes that will come
about both in the kinds of problems that will arise
and the ways in which they can be approached.

SUGGESTED READING

    The aim of this section is to suggest some
volumes which the author has found useful and
instructive. It is by no means a comprehensive list,
and those in it have been selected largely on the
grounds that they either provide a useful general
introduction to estuaries or deal with subjects
which it has not been possible to cover adequately
in this volume. More detailed and specific
references have been inserted in appropriate places
in the text, and these are included in the
Bibliography at the end.

Barnes, R.S.K. (1974) 'Estuarine Biology'. Edward
    Arnold, London.
Barnes, R.S.K. (1980) 'Coastal Lagoons: the
    natural history of a neglected habitat'.
    Cambridge University Press, Cambridge.
Barnes, R.S.K. and Green J. (eds.) (1972) 'The
    Estuarine Environment'. Applied Science, London.
Boaden, P.J.S and Seed, R. (1985) 'An Introduction
    to Coastal Ecology'. Blackie, Glasgow
Bowden, K.F. (1983) 'The Physical Oceanography of
    Coastal Waters'. John Wiley and Sons,
    Chichester.
Clark, R.B. (1986) 'Marine Pollution'. Clarendon,
    Oxford.
Clark, R.B. (1987) 'The Waters around the British
    Isles: their conflicting uses'. Clarendon,
    Oxford.
Cronin, L.E. (ed.) (1975) 'Estuarine Research (2
    vols)'. Academic Press, New York.
Curds, C.R. and Hawkes, H.A. (1983) 'Ecological
    Aspects of Wastewater Treatment'. 2 Vols,
    Academic Press, London.
Dyer, K.R. (1986) 'Coastal and Estuarine Sediment
    Dynamics'. John Wiley and Sons, Chichester.
Gray, N.F. (1988) 'The Biology of Wastewater
    Treatment'. Oxford University Press, Oxford.
Head, P.C. (1985) 'Practical Estuarine Chemistry'.
    Cambridge University Press, Cambridge.
Kennish, M.J. (1986) 'Ecology of Estuaries. Volume

1. Physical and Chemical Aspects'. CRC Press Inc., Florida.

Kinne, O. (1970-1986) (ed.) 'Marine Ecology. (5 vols)'. John Wiley and Sons, Chichester.

Knox, G.A. (1986) 'Estuarine Ecosystems. (2 vols).' CRC Press Inc., Florida.

Long, S.P. and Mason, C.F. (1983) 'Saltmarsh Ecology'. Blackie, Glasgow.

McLusky, D.S. (1981) 'The Estuarine Ecosystem'. Blackie, Glasgow.

Mance, G. (1987) 'Pollution Threat of Heavy Metals in the Aquatic Environment'. Elsevier Applied Science, London.

Mann, K.H. (1982) 'The Ecology of Coastal Waters'. Blackwell, Oxford.

Meadows, P.S. and Campbell, J.I. (1987). 'An Introduction to Marine Science'. 2nd Edition, Blackie, Glasgow.

O'Kane, J.P. (1980) 'Estuarine Water Quality Management'. Pitman Advanced Publishing Program, Boston.

Olausson, E. and Cato, I. (eds.) (1980) 'Chemistry and Biogeochemistry of Estuaries'. John Wiley and Sons, Chichester.

Perkins, E.J. (1974) 'The Biology of Estuaries and Coastal Waters'. Academic Press, London.

Phillips, D.J.H. (1980) 'Quantitative Aquatic Biological Indicators'. Applied Science Publishers, Barking, Essex.

Riley, J.P. and Chester, R. (1978) 'Chemical Oceanography'. Academic Press, London.

Wiley, M.L. (ed.) (1976 - 7) 'Estuarine Processes'. 2 Vols., Academic Press, New York.

Wiley, M.L. (ed.) 'Estuarine Interactions'. Academic Press, New York.

Wilson, J.G. and Halcrow, W. (eds.) (1985) 'Estuarine Management and Quality Assessment'. Plenum Press, London.

Chapter One

THE ESTUARY AS A RESOURCE

1.1 HISTORY OF ESTUARINE USAGE

Man has always used the estuary as a resource: initially for food, then for transport and communication and most recently as a site for industry and development. Primitive man was probably attracted to the estuary by the abundant supply of food such as birds, fish and shellfish, and evidence of the part they played in his diet can be found in the excavations of prehistoric middens.

Such uses can have had little effect on the estuary, the more so since the supply of some of the food must have been seasonal, with salmon coming up the rivers in the spring and the birds coming in the winter. Likewise, the hunters must have timed their visits with the seasons, and the number of permanent residents and users of the estuary would have been quite small.

It was not until permanent settlements began to be set up on the estuaries that the character of the estuary itself started to change. The initial settlements would have been established on the most convenient point of the estuary, that is where it was still possible to cross it on foot or horseback. Most of the major towns and cities on estuaries began in this way, and even today many of them show evidence of their ancestry in their name. For example there are Deptford (literally "deep ford") and Catford on the Ravensbourne River which enters the Thames as Deptford Creek at Greenwich Reach on the Thames, and the Irish name for Dublin is Baile Atha Cliath - "the place of the hurdle ford". From this base, the settlement expanded, modifying and in places obliterating the estuary as it grew . This evolution of a "typical" European-type estuary is shown graphically in Figure 1.1 with the different

9

Figure 1.1   Changes in a European-type estuary,
              showing the growth of the settlement
              and the alteration in estuarine
              configuration. Stage A - initial
              settlement at lowest crossing point;
              Stage B - construction of sea-walls to
              maintain depth of channel; Stage C -
              infilling of intertidal mudflats and
              saltmarsh and spread of urbanisation
              seaward and down the coast.

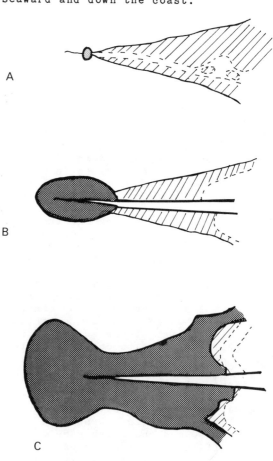

A

B

C

phases of development and utilisation.

The spread seawards was initiated by the growing use of the estuary for transport and communication, a usage which has remained of primary importance to this day. As ships got bigger, they required deeper and deeper water, and so the quayside was extended down the estuary into deeper and deeper water. The mudflats or saltmarshes behind the quays were then filled in, either by sedimentation and accretion or by man's intervention by dumping and infilling and the new land could be used for building. In Louisiana, coastal wetlands are being lost at a rate of approximately 50 square miles a year and thousands of acres have been lost from other coastal States (EPA, 1987).

This loss of mudflat and saltmarsh, which before had been the areas of deposition of silt in the estuary sometimes led to the new deeper channel in turn becoming silted up, and measures had to be taken to keep it clear. These included dredging of the channel, which was expensive and had to be repeated every few years, or the deliberate constriction of the mouth of the harbour so that the water currents themselves would keep the channel clear by scouring. Figure 1.1 shows the effect of this process on the "typical" urbanised estuary. One point to note is the different time-scales of the change in different parts of the world. In Europe, the changes have been spread over centuries, although the pace of change has been increasing all the time. Porter (1973) chronicled the explosive growth in both size and manufacturing capacity of four industrialised British estuaries that ocurred during the nineteenth century.

In the city of Dublin, for example, the remains of the Viking quays from the tenth century have been found well up the estuary, near the centre of the old walled city (Figure 1.1A). The long harbour walls (Figure 1.1B) were constructed at the end of the eighteenth century, and the past fifty years or so have seen considerable infilling and building on what were previously intertidal areas both inside and outside the sea walls (Figure 1.1C).

In other estuaries in which development owed more to industrialisation, and in many of the estuaries in the New World, the major expansion of estuarine usage ocurred rather later, although the general pattern remains the same. Today, over 70% of the population of the United States lives in the coastal states, and it is predicted that by the beginning of the 1990's over 75% will live within 50 miles of the coast (EPA, 1987).

Because of the advantages of estuaries and
shipping in the transport particularly of large
quantities of materials, many industries established
themselves there. In turn, these industries have
changed the estuary, but in a more insidious way.
The estuary had always been used as a receptacle for
the wastes of the towns or cities on it, and the new
industries simply continued the practice. Many of
the problems facing estuaries today are due to this
cavalier solution to the problem of waste disposal.

As the water quality in the estuaries
deteriorated, the traditional uses of the estuary
such as fishing and shellfishing disappeared, and
the estuary became little more than a passive agent
for transport. The disappearance of these
traditional uses was due to two reasons. The first
was that in many cases the fish themselves
disappeared, with the result that many of the major
estuaries had lost fish such as the salmon well
before the end of the last century. The second
reason was that even those hardier species that
could survive became contaminated; with shellfish
such as mussels, with their powers to concentrate
pathogens, being implicated as causative agents in
outbreaks of typhoid and cholera. Once this had been
realised, the collection of shellfish from the
estuaries was banned by the authorities, although to
this day some of the estuaries themselves still
support flourishing populations of shellfish (West
et al., 1979).

By now the advent of iron and steel ships and
steam propulsion had dramatically increased the size
of vessels needing to use the port facilities. For
example, during the so-called "Golden Age" of sail
with the tea clippers from China the vessels were
comparatively small, about 1,000 tons in size.
However, Brunels's "Great Britain", which pioneered
the widespread use of iron as a shipbuilding
material, was built thirty years before clippers
such as the "Cutty Sark" were making their
record-breaking runs, and at 3,270 tons, was nearly
three times the size of the largest wooden vessel.

To cope with the larger ships the port
facilities moved seaward, and likewise new
shipbuilding industries, needed for the new breed of
metal ships moved seaward into the now reclaimed
flats and saltmarsh in search of deeper water.

A comparison of almost any of the industrialised
or settled estuaries with its original form (see the
hypothetical example in Figure 1.1) will show that

12

a large proportion of the intertidal area has been lost, and this process is a continuing one to this day (Porter, 1973).

The physical demands on the estuary, in terms of the size of ships were to grow much more slowly during the next hundred years or so. Brunel's successor to the "Great Britain", the "Great Eastern" launched in 1855 weighed almost 20,000 tons (at which size she proved almost too big to be launched from her yard at all) and she may be compared to the liner "United States", which had a displacement of 53,000 tons and made her Blue Riband crossing of the Atlantic in 1952.

Shortly after this there was another dramatic leap in size with the advent of the supertanker and the bulk carrier. The largest of these were again an order of magnitude bigger that anything that had gone before, and likewise demanded facilities beyond the scope of most existing ports. Few estuaries however could cope with these new demands, and either completely new sites were chosen, with deep water and plenty of manoeuvering space being the priority rather than shelter or market access, or yet another seaward move this time right to the mouth of the estuary was made.

A more noticeable impact on the usage of the estuary was made by the shift in cargo transport to containerisation. This was intended to minimise the turn-round time of the ship by speeding up the loading and unloading times. It also demanded different handling and storage facilities, as well as requiring much less labour than the traditional methods. In many cases it was easier to build a new port altogether, again usually further down the estuary toward the sea, and this, combined with a general decline in the use of shipping, particularly for local freight or passenger services, has meant large areas of derelict, neglected and crumbling dockland in many cities.

Industry too has been changing, with many of the traditional heavy industries or those such as papermaking or tanning becoming more efficient (i.e. less wasteful) or contracting or disappearing altogether (Mackay et al., 1978). The new industries which have taken their place have usually been designed with some thought to the environment, and in general it is still the older industies and the older factories which pose the problems. However, the new industries are not without their own particular problems, notably in the substances which they discharge. The most obvious examples of these new discharges are radioactivity which dates back only to the 1950's and chemicals of which new examples are being developed daily.

Widespread public concern for the environment in its own right is a relatively recent phenomenon, and as mentioned above it has already had an impact on the usage of the estuary through the design or operation of industries. Nevertheless, the notion that there is an amenity value in economic terms for such areas is still new and even now not always accepted. It can however be made much more acceptable if it can be shown that protection of the estuary for amenity purposes also aids economic purposes such as fishing! Amenity or recreational usage is a growing demand, especially in industrialised countries where such areas may be at a premium, and like all trends has to be taken into account in the management process.

1.2 TYPES AND VALUES OF USAGE

Each estuary has its own particular pattern of use, which has evolved over the years according to the demands made on it. As an example Figure 1.2 shows the values of the different commercial usages of a fairly typical estuary, in this case the Clyde estuary and Firth of Clyde in Scotland. The total value is around £40 million, of which the major part is waste disposal. This figure for waste disposal is perhaps somewhat inflated, as it includes the costs of all waste treatments in the catchment area, and not just those which go directly into the estuary. Nevertheless, since all the waste reaches, or would reach the estuary in the end, this figure is probably the most realistic.

In contrast, the figure for fishing represents only the commercial catches, and does not include the value of sport fishing catches. Since substantial salmon fisheries exist on several of the rivers, this could amount to quite a sum. However it is difficult to cost this on market value of fish caught alone, as much of the value of such a fishery is in the associated costs (rod fees, accommodation etc.), and these may add to the value by as much as two orders of magnitude per fish caught.

The value of the estuary as an amenity is clearly shown in Figure 1.2 and in fact exceeds the strictly commercial value (shipping and fishing). At peak times, there may be over 160,000 visitors to the area (NERC, 1974) and, even assuming a

14

conservative estimate of spending for each, this
makes tourism a major economic factor for the area.
   The shipping value, representing the value of
goods carried, reflects the decline in industry of
the area in general. Traditional heavy industries on
the Clyde such as steel or shipbuilding itself have
contracted down to a fraction of what they were at
their peak. The major contribution to the figure for
shipping is oil, which accounts for over 50% of the
total tonnage shipped. Iron ore is still important,
and is the next major cargo item at around 17% of
the total tonnage.

Figure 1.2   Proportional value of the different uses
             of the Clyde estuary and Firth of Clyde
             in Scotland expressed as a percentage of
             the total budget. (Data from several
             sources).

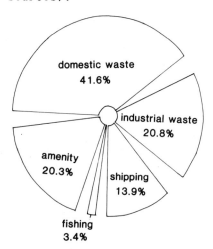

   The value of the estuary directly to industry is
difficult to estimate. The water is used as cooling
water in power stations, and of course, the costs
of waste treatment, excluding this thermal effluent,
are already shown. In that such thermal effluents
seem to have little effect, and therefore generate
no costs, they have been excluded from Figure 1.2.
In other areas the estuary may also have a direct
industrial value as an extractive resource, either
as a source of water for desalination or as a source
of sand or gravel. However locally important these

may be, they are of minor importance in estuaries as a whole.

Each of these main categories, transport, waste disposal, fishing and amenity can be broken down to see if it is possible to identify trends to be considered for future management strategies.

As mentioned above there have been rapid changes in the modes of shipping. The success of the internal combustion engine has led to the great increase in roll-on roll-off (RORO) traffic. This increase has been as marked in passenger transport as it has in freight and the result has been the same, with a change in the kinds of facilities needed (Clark, 1987).

1.3 SHIPPING

The present pattern of merchant shipping is shown in Figure 1.3. This summary, which represents the data for the U.K. as a whole, is close in general terms to the example for the Clyde discussed above. Again the major item is oil, with other bulk cargo such as ores of lesser importance. If these figures were to be compared with earlier data, the most obvious differences would be a decline in conventional shipping, by about 50% in the past ten years, and a doubling of the RORO traffic in the same period. The Clyde estuary is able to handle large ships such as supertankers and bulk carriers, but this may not be the case in all estuaries.

Nevertheless, the size of ships to be handled will be of prime importance in management. The costs of dredging to keep the shipping channel clear are considerable, and in addition there is the problem of disposal of dredge spoil. This problem can be complicated by the pollution status of the spoils, especially from heavily industrialised estuaries.

In addition oil from shipping does pose particular problems in terms of management and pollution control. Catastrophic oil spills such as the wreck of the "Torrey Canyon" off Land's End, or the explosion which blew up the "Betelgeuse" in Bantry Bay in Ireland are fortunately rare, and the majority of spills are small. Although the introduction of the LOT (Load On Top) system has greatly reduced the amount of oil released into the marine environment by shipping, the use of any estuary by ships, and not just oil tankers will mean some oil in the estuary. The greater the volume of oil, fuel or otherwise, handled, the greater the amount that is likely to end up in the estuary.

Figure 1.3 Proportional values of U.K. shipping.

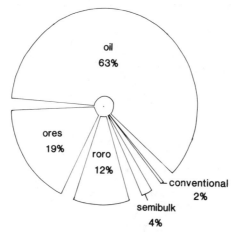

1.4 DOMESTIC AND INDUSTRIAL WASTES

The wastes loading on the estuary will depend on the volumes and types of the discharges into it. While domestic effluent load may be estimated with a reasonable amount of accuracy from a knowledge of the population in the catchment area, industrial wastes are less easy to calculate. The kinds of wastes associated with a particular industry can usually be ascertained without much difficulty, but the problem arises in gauging the quantity released, which will vary with the efficiency of the processes (which in turn often means with the age of the plant). The kind of treatment needed will also vary with the type of effluent, and in many cases the most efficient solution is to treat industrial effluents separately from the domestic. Again this may only be possible where new industries have been introduced, and in the majority of cases, the industrial effluent has been discharged into the nearest sewer leading to problems not only of treatment of the mixed effluent, but also of tracing pollution incidents to their source.

As with the other demands on the estuary, the wastes loadings have changed over the years. Figure 1.4 shows how the load on the River Thames has grown with the increase in size of the city. Before 1830

17

and the introduction of the water closet, the discharges into the river would have been through a large number of diffuse sources from the thousands of cesspools in the city, although this did not stop the Bishop of London complaining as early as 1620 that it was time something was done about the river. Despite the implication of comments like these, the River Thames continued to yield large catches of salmon, with records of up to 130 landed in one day. However by 1833 the last salmon had gone and by 1850 all commercial fishing, which had kept as many as 400 in employment between Deptford and London, had ceased.

Figure 1.4   Changes in BOD load on the Thames: A - advent of water closet; B - commissioning of chemical precipit- ation plants; C - introduction of widespread secondary treatment. (Data recalculated from several sources).

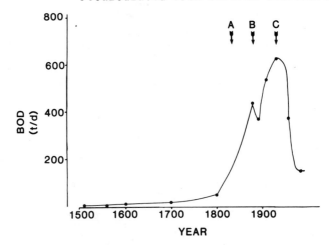

A more serious problem was the threat to public health, and by the middle of the nineteenth century there had been enough of a link established between contaminated water and the epidemics for legislation to be brought in to ensure supplies were drawn fron clean water above the discharges. As an aid to this process, outfalls were also constructed to take the sewage beyond the metropolitan limits.

However as Figure 1.4 shows, London was growing at a tremendous rate, and by 1880 the population was increasing by about one million people every ten years. The industrial load on the estuary had also been growing throughout the nineteenth century, notably from coal gas effluents, including by 1880 those from the largest works in Europe. Although the direct industrial load on the estuary was slight, some 10% of the total load, indirect industrial loads through the sewerage system accounted for almost one-third of the input into the estuary.

The condition of the Thames not unnaturally worsened under this pressure, and the 1890's saw the introduction of treatment, in the form of chemical precipitation by ferrous sulphate, for the north and south outfalls. This action did succeed in improving things at least for a while, (see Figure 1.4) and fish did return to parts of the river and estuary from which they had been absent for years. However, this respite was temporary as the city continued to grow, albeit more slowly after 1910, and it was not until the advent of further treatment works including biological treatment in the 1950's and the commissioning of comprehensive treatment plants in the 1960's and 70's that significant improvements were made. The improvement is so marked that now almost one hundred different species of fish have been reported to be in previously dead stretches.

The growth of heavy industry, notably iron and steel making, made major contributions to the pollution loads, not only through the wastes discharges in such processes, but also in the increased mobilisation of metals from the mining activities. Analysis of sediment cores from the North Sea and from other locations such as the Severn has shown a picture as depicted in Figure 1.5, where the amount of contamination has increased several-fold in the past one hundred and fifty years or so (Allen and Rae, 1986).

In many cases, particularly in Britain, the development followed a similar pattern (Porter, 1973), and the principles and practices as well as the system laid down in the latter half of the nineteenth century still form the basis of the system today. The Royal Commission on Sewage Disposal, set up to examine the problem of pollution in estuaries and tidal waters reported its findings in 1908, and recommended guidelines for such discharges. These guidelines stressed dilution of the effluent rather than treatment, and it is only relatively recently that there has been an acknowledgement that the capacity to dilute and absorb was not infinite and that some sort of treatment to lessen the burden on the system was necessary.

Figure 1.5 Industrial loading on the Severn estuary, U.K., showing relative increase in lead (————), copper (— — —) and Zn (······) with time (from Allen and Rae, 1986).

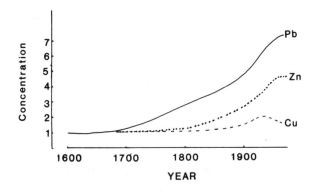

1.5 FISHING

The Victorians also demonstrated the link between sewage pollution and outbreaks of such diseases as cholera or typhoid; this link was also noted via the consumption of shellfish. Many estuaries used to have flourishing shellfisheries, which have now been lost, either through the loss of the beds themselves through pollution or over-exploitation (even in unpolluted estuaries) or through a ban for public health reasons on collection and sale.

Nevertheless, fishing does remain an important part of the economy, particularly the local economy, in many estuaries (see the Clyde example in Figure 1.2). In addition, in the Third World fish are a vital source of protein, yet in places local fisheries have been lost due to pollution. The total value of fish landed at British ports in 1984 was almost £300 million, although it should be noted that of this sum relatively little was actually caught as opposed to being simply landed in estuaries. A better example of the potential of estuaries for fishing is shown in Figure 1.6, which gives the values of the different catches from the Potomac estuary on the east coast of the United

20

States. This is a famous fishery, especially for the
Blue Crab, which is a delicacy and attracts a
premium price when sold. Estuaries are particularly
good places for shellfish as the example in Figure
1.6 shows. Shellfish account for over 60% of the
value of the catch compared to a percentage of 16%
(of which the deeper water Norway lobster or Dublin
Bay prawn accounted for over 6%) in the total
landings for Britain.

Figure 1.6   Proportional value of fish catches in
             the Potomac estuary, USA, expressed as a
             percentage of the total.

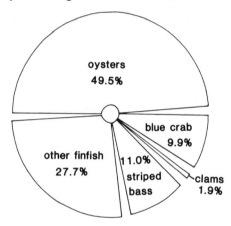

Commercial landings of fish in the United States
in 1985 had a dockside value of some 2.3 billion
dollars, and a retail value several times that, and
fish and shellfish harvested from coastal and
estuarine waters (within three miles from shore)
accounted for roughly half this revenue. In
addition, estuaries and coastal areas are of prime
importance as breeding and nursery grounds for the
juvenile stages of around two-thirds of all
commercial species (EPA, 1987).

Many estuaries and coastal areas are now being
increasingly utilised for mariculture, with the two
main areas of activity being shellfish, primarily
mussels and oysters, and salmonids. In many European
estuaries, shellfish culture is a traditional and
long-established practice, and such a use has been
incorporated into the management of the estuary as

it has developed. In these situations, attention has
always been paid to the quality of the estuary, as
the contribution on mariculture to the local economy
can be considerable.

In this context it is worth noting the speed of
the response of the French and American Governments
to the danger posed by anti-fouling paints
containing TBT (tri-butyl tin). It had been noticed
that oysters in particular, cultivated near harbours
or marinas, were growing abnormally, meat yields
were down, and spat settlement was poor, and this
situation was quickly linked to the new paints which
had only recently appeared on the market. In both of
the above countries, mariculture and shellfisheries
are of great importance and the use and sale of TBT
was quickly restricted. In contrast, in the UK where
such industries are much less important, similar
legislation lagged some years behind that of the
example given by France and the USA.

As with the Clyde example, sportfishing has been
excluded from Figure 1.6. Again though, the
contribution of sportfisheries would be
considerable, and in the Potomac system, the sport
catch of species such as Striped Bass may be up to
25% of the commercial landings. Adding on the extra
value to these catches could add half as much again
to the total value of the fishery.

1.6 AMENITY

The amenity value, which includes the
conservation value has always been difficult to
quantify in strictly economic terms, although
considerable weight has always been given to
considerations such as sport fishing, especially for
prestige species such as the salmon.

It has been estimated that the authorities in
the USA spent over $5 billion in 1982 on coastal
recreation facilities, and these areas received some
60 million visitors in 1985, including 22 million to
the seashore alone (EPA, 1987).

Figure 1.7 shows the amenity usage of Dublin Bay
in Ireland, and shows the range of leisure interests
which must be considered. The total numbers of
people involved are only some 6% of the Dublin area
population, but even at that level there are
problems of over-use or under-supply for some of
these activities. Some idea of the value of these
activities in the local economy can be gained from
the example of the sailing activities. There are

around 20,000 people who use the Bay for sailing,
and even assuming only some 5% actually own and run
a boat, this still amounts to an annual expenditure
of one million pounds on running costs alone.
Indeed, a survey of employment in the area
(O'Sullivan, 1987) showed that there were more
people employed in the pleasure boat industries and
services than in the international ferry and
mailboat terminal nearby.

Figure 1.7   Proportional amenity usage of Dublin
Bay, Ireland. (From O'Sullivan, 1987).

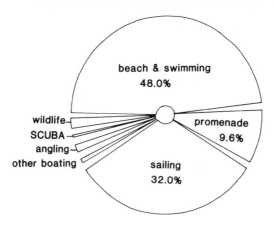

The demand for watersports is growing, and
activities such as SCUBA diving or sailboarding are
uses which could not have been foreseen twenty years
ago. Because these sports bring their participants
into fairly close contact with the water, there is a
greater chance of contact with potential pathogens
and also a greater likelihood that the participants
will be concerned for the quality of the water. For
these reasons, the public health aspects are being
scrutinised more closely and also the pressure to do
something about it is growing.
    Although birdwatching or observation of wildlife
may only make up a relatively small proportion of
the leisure usage of the estuary (see Figure 1.7),
this particular function is of prime importance.
Apart from the commercially-related conservation
aspects, such as nursery grounds for juvenile fish

and especially flatfish, estuaries are also valuable refuges for wildlife. This importance is often the more marked because the estuary may be the only large wild (i.e. unmanaged) area near the city, and as such offers an unmatched opportunity for observation and study. In Dublin Bay this importance has been recognised nationally and internationally by the declaration of Bull Island as a UNESCO Biosphere reserve in 1981. Bull Island, although showing some signs of pollution from the surrounding city of Dublin is nevertheless of prime importance for birdlife. A total of 180 species have been recorded there, but its major value is as the winter haunt for up to 40,000 wildfowl and wading birds.

The importance of estuaries for birdlife is shown in Table 1.1. These examples are from the Shannon estuary on the west coast of Ireland, which is a rare example of a large unpolluted European estuary, and from Bull Island itself.

Table 1.1   Waterfowl in the Shannon estuary and Bull Island of national (*) or international (**) importance. (Based on Merne, 1985)

Species	Qualifying numbers		Numbers	
	International	National	Shannon	Bull Is.
Greylag Goose	900	50	68*	<1
Brent Goose	200	200	60	1500**
Shelduck	1250	750	852*	800*
Widgeon	5000	2000	7816**	2600*
Teal	2000	1000	2735**	1200*
Shoveler	1000	90	234*	300*
Scaup	1500	100	273*	<20
Pintail	750	250	16	300*
Dunlin	20000	5500	25490**	7900*
Knot	3500	2500	900	6700**
Black-tailed Godwit	400	50	5515**	300*
Bar-tailed Godwit	5500	450	929*	2300*
Curlew	3000	1000	3024**	1900*
Redshank	2000	1000	3383**	2400**
Oystercatcher	7500	3000	230	3800*
Grey Plover	800	100	24	150*

In addition to the eleven species shown above, the Shannon estuary is of prime importance to other species such as Lapwing and Golden Plover, whose numbers fluctuate erratically but on occasion exceed 5000, and Ringed Plover whose numbers like those of the Pintail, have declined below the qualifying levels in recent years.

The value of estuaries to birdlife is threatened by a variety of changes. Firstly there is the direct threat of habitat change or loss, for example when the mudflats are infilled in waste disposal. This tendency to regard mudflat and saltmarsh as "wasteland" with no value which can be "reclaimed" for use is still prevalent, but it should be noted that these areas play a considerable part in many of the estuarine processes (see Chapter 2) apart from their conservation role.

Secondly, the increasing use of the estuary means more disturbance. This can affect not only birds which nest on the estuary but also those which use it for roosting or feeding. Overwintering birds are particularly vulnerable to this sort of pressure and any additional stress can significantly add to the mortality rate.

Thirdly, there is the problem of pollution, although sometimes the evidence is conflicting. For example the slightly-polluted Bull Island has bird densities (measured as numbers per hectare of mudflat) several times higher than the unpolluted Shannon, but it is those cases where there are persistent pollutants such as metals or organochlorines that present the greatest threat. Organochlorine residues off the coast of Holland, where the Wadden Sea is arguably the most important area in Europe for birds, are about two orders of magnitude greater than those of baseline areas (Wilson and Earley, 1985). While the bird populations of the Wadden Sea seem to be holding their own, other vertebrates such as the seals are declining, and the causes and possible remedies for the decline are being intensively investigated (Klinowska, 1986).

Conservation is an estuarine usage which conflicts with every other use, and epitomises the management dilemma. How much disturbance should be allowed? Should it be restricted to scientific disturbance for the purpose of census and stock evaluation, or should birdwatching by the public, which is important both in terms of numbers of people involved and in the emotive power of the issue, be permitted albeit in a strictly controlled fashion? In the light of the location of many estuaries close to centres of population, they offer excellent facilities for education at all levels.

Not least of the problems of access is that of litter left by visitors, and again in public perception this is one of the most important criteria for quality judgement. Although a lot of

litter on beaches is of marine origin, dumped from
ships in defiance of the London Dumping Convention
(see Chapter Five), considerable quantities are
discarded by the very people using the estuary or
the beach for recreation (fishing, picnicking).

On the wider scale there are conflicts also
between the other estuarine usages, and these are
summarised in tabular form in Figure 1.8. The
various kinds of usage can be grouped into two
categories: those which are system dependent, and
those which are not. Those which are system
dependent rely on the more or less natural
functioning of the estuarine ecosystem, and if the
system breaks down or is degraded, then these usages
are likewise impaired. The usages that are system
independent do not require a functioning ecosytem,
and therefore quality is of little importance. The
use of the estuary to receive waste can belong to
either category. If physical dilution is the sole
aim, then obviously it is system independent, but if
some measure of waste transformation or breakdown is
intended, then this will be system dependent.

The major conflict is between these two main
categories, and within each category there is a
rough compatibility. Conservation presents one of
the most intractable problems, as the provision of
totally untouched areas is firstly in total
opposition to the idea of management, and secondly
at odds with multiple optimal usage of the resource.

The usages of the estuary and the various
interests involved will be the prime factor in
setting the EQO or Environmental Quality Objective,
which forms the basis of the management plan for the
estuary and this will be discussed in a later
Chapter.

The estuary as a resource

Usage	System Independent				System Dependent			
	Shipping	Indust.	Extract.	Waste	Fish.	Maricult.	Amenity	Conserv.
Shipping	–	*	*	*	**	**	**	**
Industrial		–	*	*	**	**	**	**
Extraction			–	*	**	**	**	**
Waste				–	***	***	***	***
Fishing					–	*	*	**
Mariculture						–	*	**
Amenity							–	*
Conservation								–

Figure 1.8 Conflicts of usage in an estuary; the more stars the greater the incompatibility between uses.

27

Chapter Two

THE ESTUARINE SYSTEM

2.1 ESTUARINE CLASSIFICATION

In the previous chapter, the influence of man's activities on the estuary was outlined, and we shall now go on to look at the underlying estuarine system. This system can be thought of as three interacting components, namely the physical, chemical and biological sub-systems, and it is these which we will now go on to consider.

2.1.1 Physical classification

As has already been mentioned, the term "estuary", for the purposes of this volume will also include most other coastal situations such as bays, lagoons and deltas. The term "estuary" itself is in any case difficult to define, and may mean different things to different people. Cameron and Pritchard (1963) defined it principally in terms of physical structure or topography -

"An estuary is a semi-enclosed coastal body of water which has a free connection with the open sea and within which sea water is measurably diluted with fresh water derived from land drainage"

This definition provides a good working basis, but clearly is wide enough to cover most of the other situations above, depending on the degree of enclosure and the degree of dilution. The physical topography of the estuary has been used by Pritchard (1955) as the basis of a simple system of classification into the four types described below, and the three main types are shown in Figure 2.1.

a) Coastal plain estuaries

These are what are often thought of as typical estuaries, and are the commonest types especially in northern temperate waters. They are formed by the flooding of river valleys, and their physical shape retains the characteristic "V" shape of the parent valley. Although sedimentation is relatively low, and this is why the original shape is retained, these estuaries are usually characterised by extensive mudflats and saltmarshes. When the tide is out, the underlying topography of flat mud cut by channels can be seen. The channels have a high depth/width ratio, but the flats are the reverse, and this duality has to kept in mind when considering other processes such as mixing.

Figure 2.1   Estuarine types: A) Coastal plain estuary; note the complex and shifting series of channels: B) bar estuary: and C) fjord - note the limited intertidal area (indicated by the dotted line).

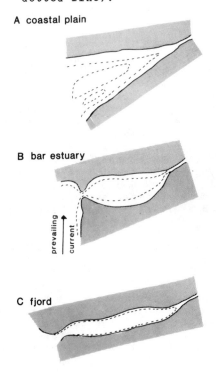

A coastal plain

B bar estuary

prevailing current

C fjord

Because of their shape, widening out as they approach the sea, these estuaries grade into the marine environment proper and consequently it is difficult to say where the estuary ends and the sea begins. For practical purposes of management, it is easier in many cases to include the bay into which the estuary is discharging as part of the same system, and to include it in such exercises as impact assessments and mathematical modelling.

b) Bar estuaries

Bar estuaries are similar in many respects to the coastal plain estuaries described above, with the major difference being the greater sedimentation in bar estuaries. As a consequence, a bar is deposited across the mouth of the estuary where the estuarine and marine sediment transport processes cancel out. These estuaries are more usually found in tropical regimes, but can be found on any coastline of active deposition.

The bar across the mouth severely restricts the area of interchange between sea and estuary, and often provides a convenient conceptual as well as physical boundary between the two systems. Within the bar, the estuary displays a certain consistency of form, and although it is seldom more than a few metres deep and so has a low depth/width ratio, it lacks the complexity of channel and flat of the coastal plain estuaries.

Bar estuaries often have extensive shallow lagoons behind the bar and these lagoons, which can be either hyper- or hypo-saline depending on conditions, often develop unique and highly individual biological communities. The development of such populations is possible because the lagoons, although subjecting the biota to considerable environmental stress, do not show the same temporal variability of conditions exhibited by the mudflats and saltmarshes of the coastal plain estuaries.

c) Fjords

Fjords, like the previous two estuarine types, are river valleys flooded by the sea, but in this case the valley is deepened and widened by the action of overlying ice prior to flooding. The deepening and widening results in the characteristic rectangular cross-section of fjords, and this is combined with a very high depth/width ratio. In addition, many fjords have rock sills at the mouth, with the result that although the fjord itself may be some hundreds of metres in depth, the depth at

the mouth may be only tens of metres or less. As
with the bar estuaries, exchange between the sea and
the estuary is restricted, but because of the
greater depth of the fjords compared with the bar
estuaries, the fjords have a much greater tendency
toward stratification and to anoxia in the lower
layers.

Because fjords are restricted to mountainous
coastlines, their catchment area for river input is
restricted, and due to the small riverine input, and
to the geomorphology of the river courses, the
amount of sedimentation is very slight.

Likewise the nutrient input, in the absence of
anthropogenic inputs, is slight, and the depth and
stratification of fjords mean that the nutrient
cycle and energy flow are fundamentally different
from shallow estuaries, being water column rather
than sediment based.

d) Other types
This group includes those few estuaries which do
not fit into the first three categories outlined
above. Into this category fall those estuaries which
have been formed by local geological phenomena such
as landslips or volcanic activity. Although their
physical shape may not fit them into any regular
classification, the actual processes or systems will
operate according to the same principles.

The above scheme gives a good basis for the
classification of estuaries, but it will quickly
become obvious in practice that estuaries do not
fall easily into such discrete categories, but
rather that there is a continuum of types from one
to the other.
This idea of a continuum of types along a
gradient of conditions has been put forward by
Jeffrey et al (1980), whose classification scheme,
based on the familiar gravel/sand/silt sediment
diagram is shown in Figure 2.2. This scheme
classifies the estuaries according to the numbers of
fluvial inputs, in other words whether it is one
discrete system or a number of joined systems, the
predominant sediment type, indicating the degree of
water movement and the extent of intertidal area as
well. Thus the typical coastal plain estuary might
emerge as a Type 2a11 estuary on Figure 2.2, with a
single fluvial input, predominantly sandy sediment
and over 30% of its area exposed at low tide.
The physical, chemical and biological processes
all depend to a greater or lesser extent on the

The estuarine system

topography of the estuary and so it should also be
possible to define estuarine types according to
these parameters.

Figure   2.2   Estuarine classification scheme of
               Jeffrey et al. (1980).

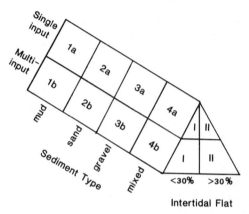

Intertidal Flat

    The dominant physical process in the estuary,
certainly in terms of its effect on the other
processes, is the mixing of the fresh and salt
waters. As mentioned above, the tendency of an
estuary to stratify into layers of water of
different densities depends on the density
differences of the constituent water bodies, the
ratio of depth to width and also the degree of
disturbance of the water body. This disturbance can
be wind generated, in which case the degree of
exposure and the fetch of wind on the water will be
dominant factors, or by a combination of bottom and
current generation, for example by tidal current
entering over a shallow sill or through a narrow bar
mouth.
    The effect of width and depth on the behaviour
of the water bodies in the estuary is depicted
diagrammatically in Figure 2.3. In the estuary
freshwater meets denser salt water; in a
frictionless system, these two waters would remain
essentially discrete (apart from diffusion mixing),
and the freshwater would continue to flow seaward on
top of the salt water. This is essentially the

32

position shown in Figure 2.3A, where the fresh and
salt waters are separated by a thin layer of mixed
or brackish water. In shallower situations mixing
becomes more and more prevalent until the completely
mixed stage of Figure 2.3B is reached. In larger
estuaries another force, the Coriolis force due to
the earth's rotation comes into play by displacing
the water movements to one side or another. In the
northern hemisphere, movement is deflected to the
right, and in the southern movement is deflected to
the left. The effect of this in a large estuary is
shown in Figure 2.3C, where the two banks of the
estuary are in fact subject to different regimes.

Figure 2.3 Effect of increasing width and
decreasing depth on water
stratification and circulation in an
estuary (after Pritchard, 1955).

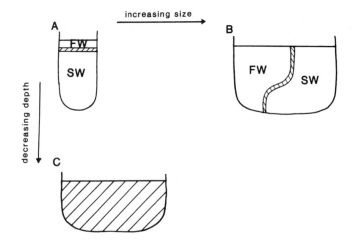

Fjords are extreme examples of type 2.3A. Mixing
is restricted to a short zone adjacent to the sill,
whose presence, although increasing local mixing,
nevertheless restricts the total water movement in
and out, and freshwater input at the head is also
limited. Lagoons, on the other hand more closely
approach type 2.3B.
     In longitudinal section (Figure 2.4) these
differences in water circulation and mixing between
estuaries, lagoons and fjords can be more clearly
seen. All three have an input of freshwater at the

head, inputs and outputs (with the tide) of salt water at the mouth and also an output of freshwater through evaporation from the water surface. In an estuary (Figure 2.4A), there is a mixing zone in the middle regions which may move up and down with the tide, and the haloclines between the fresh and sea water can assume any inclination from the near-vertical to the near-horizontal depending on current speeds and the length of the estuary.

Figure 2.4   Water circulation and mixing: A) estuary; B) lagoon; and C) fjord.

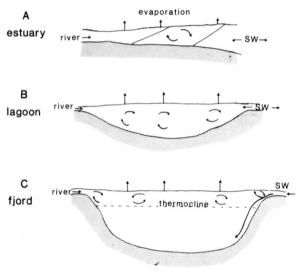

The lagoon (Figure 2.4B), because of its shallow depth and large surface area, is well mixed throughout. There is a salinity gradient between the freshwater and saltwater, but this is a gradual change over the length of the lagoon without the abrupt changes that occur in the estuary. Since lagoons have a large surface area relative to their volume, evaporation can be an important factor in determining the salinity. Where the freshwater input from rivers or runoff exceeds evaporation, then the conditions in the lagoon approach those of the estuary, with salinity intermediate between fresh and salt water. When, however, evaporation exceeds freshwater input, then the lagoon becomes hypersaline and the salinity exceeds that of the seawater. The two main factors in determining the salinity are therefore the amount of rainfall and the temperature, with the result that hypersaline lagoons are largely found on tropical or semi-tropical coasts. In temperate zones, lagoons are rarely permanently hypersaline, although they may fluctuate from hyposaline in the winter, when rainfall is high and evaporation is low to hypersaline in the summer, with low rainfall and high temperatures. Where river input is particularly low, the bar at the mouth of the lagoon may close it off completely from the sea. In this case, the tendency to become hypersaline is accentuated since there is no flushing or renewal of the water in the system, and the connection to the sea is not re-opened until the autumn gales break through the bar again.

Freshwater and seawater inputs are limited by the physical shape and location of the fjords. Mixing is confined to small areas at the mouth and head of the system, and wind generated mixing of the water in the middle reaches is restricted also by the location, in which the height of the surrounding mountains provides protection against winds from almost every direction. What little mixing does occur in the middle reaches is confined to the surface layers, as in most fjords the formation of a thermocline takes place early in the year and persists right through until the winter. Even in winter, the lack of disturbance may permit layering within the water column, with very cold freshwater from small storm runoffs lying on top of warmer salt fjord water to give an inverted thermocline.

The water below the thermocline or halocline stays relatively constant throughout the year, although the lack of renewal or contact with the surface means that there is a tendency toward anoxia which may be exacerbated by man's influence in the form of organic inputs or wastes. However, the separation of the surface layers from the bottom sediments by the thermocline means that the sediment processes play a much less important role in the cycling of material such as nutrients than in conventional estuaries or lagoons, and those that sink to the bottom are then lost until the system is re-cycled by winter gales. Consequently, nutrients are limited by what is in the water column at the time of stratification, and this limitation makes a fundamental difference to the operation of the system in the fjord as opposed to the shallower lagoons and estuaries. Nevertheless, the constancy of composition of the water does make life easier for organisms than the large and often abrupt changes that characterise the other systems.

The spatial variability of estuarine waters reflects the temporal variability. Temporal variability exists on a short time-scale with the diurnal tidal cycle, on a longer time-scale with the neap and spring tidal cycle, and finally on a variable time-scale with the variations in freshwater input from river flow and runoff. The environmental conditions at any one fixed point in the estuary will change as the inputs from sea and land change, and these changes, especially the erratic fluctuations in freshwater input not only create problems for the biota, but also create practical difficulties for management, notably in sampling and the modelling of processes. They also pose problems for the meaningful comparison of estuaries, even those which can be matched within the topographical schemes above.

2.1.2 Chemical classification

Of course the estuary itself was defined at the start of the chapter not only by reference to topography, but also to the water chemistry. Cameron and Pritchard (1963) defined the seaward boundary in terms of the measurable dilution of seawater, and the landward boundary may be similarly defined, in terms of the measurable effects of saltwater. Other workers (Portmann and Wood, 1985) have put absolute limits to this by taking the landward boundary to be where the freshwater becomes unsuitable for irrigation purposes at a chloride concentration of 200mg/l.

The most widely-used classification scheme, although used more to designate different zones within a single estuary, based on salinity is the Venice system (Carriker, 1967), which recognised six distinct zones based on the mean salinity of each:

1) Limnetic or freshwater (<0.5‰)
2) Oligohaline (>0.5 to <5‰)
3) Mesohaline (>5 to <18‰)
4) Polyhaline (>18 to <30‰)
5) Euhaline (>30 to <40‰)
6) Hyperhaline (>40‰)

It is important to realise that the limit of

estuarine influence as defined chemically, may be much less than if it were to be defined physically. The physical influence of the tide propagating up estuaries into the rivers may be felt way beyond the extent of the seawater penetration and the prime example is in the River Severn in England, where the standing wave known as the Severn bore continues many miles up the river itself.

The importance of chemical processes is implicit in this alternative definition of an estuary offered by Hedgpeth (1983) -

"The estuarine system is a mixing region between sea and inland waters of such shape and depth that the net residence time of suspended matter exceeds the flushing time"

This definition is a much more process-orientated definition than Cameron and Pritchard's (1963), emphasising as it does the importance of estuaries as places where the constituents of fresh and sea waters interact and mix.

However, mixing is not neccessarily a simple process, but can be either conservative or non-conservative and the difference between these is shown diagrammatically in Figure 2.5.

Figure 2.5   Conservative mixing (solid line) of a parameter such as salinity: non-conservative mixing by addition or removal of substances is shown by the dashed lines.

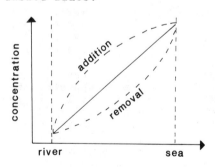

In conservative mixing, a parameter behaves

exactly as the sum of its origins; for example salinity can be predicted on the basis of the amounts of fresh- and salt-water in the water body in question. In non-conservative mixing, there is an intermediate process which alters the concentration of the parameter in question. Such processes can be breakdown processes which alter the form of the parameter, for example the breakdown of particulate organic matter, or removal processes within the system, for example the flocculation and deposition of suspended matter at the freshwater/saltwater interface.

2.1.3 Biological classification

Many biological scientists, notably Hedgepeth (1983) have criticised classification schemes such as the Venice scheme (above) on the grounds that the divisions bear little relevance to the biology of the system. However, many workers. led by Remane (e.g. Remane, 1971) have divided estuarine species into three categories, namely freshwater, brackish and marine, although there is a considerable amount of debate over the actual salinity relationships of these categories, and attempts to further sub-divide these classes have resulted in a range of rather spurious distinctions (Hedgepeth, 1983).

The scheme proposed by Odum and Copeland (1974) is radically different in that it breaks away from the reliance on salinity, treating it only as one of a number of different parameters, and instead classifying from the viewpoint of ecosystem energetics. They (Odum and Copeland, 1974) classified estuaries, along with other coastal systems, into the following categories:

1) Natural stressed systems. These are found at a range of latitudes, and the stresses may be wave action, tidal currents or salinity (hyper- and hypo-saline) itself and are characterised by low species diversity. An example would be the Lagune Madre in Texas.
2) Natural tropical systems. These have few physical stresses and are characterised by relatively high diversity and productivity - for example the tropical seagrass systems.
3) Natural temperate systems. The most frequently studied of the systems, the seasons play an important role, determining both physical and energetic structure changes.

4) Arctic systems. Here, a variety of stresses such as ice stress affect diversity and distribution, and temperature and light are major limits on productivity.
5) Emerging systems. These are those estuaries which are associated with anthropogenic loadings, and whose characteristics have been changed by this.
6) Migrating sub-systems. The final category includes those populations which are not full-time members of the estuarine system but instead use it temporarily and so act as a link between it and the other neighbouring systems.

2.2 SEDIMENT PROCESSES

The sediment in an estuary plays a major role in many of the processes, and this role is not confined to the sediment on the bottom. The availability of many items such as nutrients or heavy metals depends to a great extent on their rate of flux in and out of the sediment, and so it is worth considering the process of sedimentation and resuspension itself in a little more detail.

The movement of sediment particles is a function of the size of the particle and the current or wave movement - the water energy - over it (assuming for the moment that the particle density is constant). The relationship between the two is shown in Figure 2.6. As water energy increases, particles of larger and larger size can be moved, and as water energy decreases, progressively smaller and smaller particles settle out.

The larger particles settle out solely according to their size according to the Impact Law:

$$V = 33d^{1/2}$$

where V = settling velocity (cm/sec) and d = diameter (cm). For the smaller particles, the settling rate is determined by the viscosity of the fluid according to Stokes' Law. For quartz particles in water at $16^{\circ}C$

$$V = 8100d^2$$

where V and d are as above.

However, at the smallest particle sizes, the silts and clays, the settled particles display a cohesion binding them together and effectively requiring a much higher water energy to re-suspend them once settled.

Suspended sediment brought down in rivers tends to remain in suspension because the river channel is relatively narrow. As the river widens out into the estuary proper, the water energy is dissipated over a wider area, and the sediment begins to settle out. At the seaward end of the estuary, the water energy may increase due to tidal currents or wave action. The effect of tidal currents on sediment structure is particularly marked in bar estuaries, where one may pass from a gravel or mixed sediment in the river bed to fine muds in the estuary or lagoon, giving way to well-sorted sands or gravels at the mouth where the tidal currents sweep in and out. In coastal plain estuaries which widen out gradually toward the sea, there is a gradual increase in sediment particle size from the estuary to the fully marine beaches at the mouth, where wave motion acts to winnow out the fine particles. In areas with little tidal or wave movement, the sediment builds up around the mouth of the estuary, forming a delta.

Figure 2.6 Water energy and particle size. Settling velocity is the rate at which the particles settle (Stokes' Law) and threshold velocity is the energy required to move the particle (see also text).

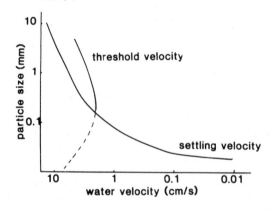

The deposition of sediment and the formation of banks or bars is affected by the course of the water currents. In large estuaries particularly, the coriolis force can deflect the water such that the flood channels of incoming seawater are deflected to one side of the estuary, and the ebb channels are deflected to the other side. The result is a complex dual system of channels on either side, often leaving a bar or banks in the middle. At the mouth

the position and orientation of the bar is
determined by the amount and direction of the water
movement, partially in the form of wave energy but
mainly as longshore drift or currents. The bar
across the mouth of the estuary points in the
direction of the residual longshore current, and the
opening into the estuary gradually moves in that
direction also, as the bar builds up and material is
eroded from the opposite shore. The resulting kink
in the shape of the estuary becomes more and more
pronounced, but all the time the water energy, which
is concentrated on the outside of the bend, is
eroding the bar. Eventually, the bar across the
mouth is broken through, and the cycle begins again.

In addition to the physical deposition of
particles described above, there is in estuaries
also the process of flocculation where the
freshwater and saltwater meet and mix. Where this
occurs, the suspended matter in the freshwater
flocculates or binds together to form larger
particles which are then in turn more likely to
settle out. As well as the salt flocculation
process, the density gradients at the interface of
fresh and salt waters concentrate the particles and
render collisions, and hence binding, between
particles more likely.

In many estuaries, this interface produces a
characteristic area of high turbidity known as the
turbidity maximum. The extent of such an area and
the concentration of suspended material in it are a
function of the estuarine circulation and the
material brought in by both river and sea.

An example of how such a process works is shown
for the Weser estuary in Germany in Figure 2.7 (note
also the pattern of salinity stratification, which,
in the Weser, is longitudinal at the head of the
estuary and vertical at the mouth). The material in
the central reaches of the estuary is continually
replenished from the riverine input brought down in
the fresh water. When this encounters the salt in
the brackish region, it flocculates and starts to
settle out depending on the density gradients. At
the same time, there is an inward flow of seawater
along the bottom, and this flow likewise bears a
load of suspended material. The seawater borne
sediment then rises up into the rest of the water
column at the zone of mixing and turbulence of the
two water bodies. The effect of this zone on both
the amount of suspended particulate material in the
water column and the composition of the bottom
sediment is clearly shown in Figure 2.7. Suspended

sediment loads of up to 70g/l have been reported,
indicating that the whole bed of the estuary has
been fluidised and brought into suspension
(Wellershaus, 1981).

Figure 2.7   Turbidity maximum in the Weser estuary:
              shading indicates successively higher
              concentrations of suspended solids,
              dotted lines indicate isohalines. From
              Wellershaus (1981).

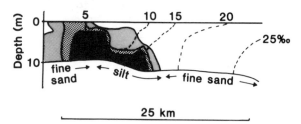

The turbidity maximum plays an important role in
the remobilisation and availability of many elements
and is particularly important in situations where
the bottom sediments have a tendency toward anoxia.
When these are brought up into the water column,
there are changes in their chemistry with oxidation
and often salinity and it is these changes which
affect speciation and mobility of substances. The
effects will become clearer after consideration of
the major pathways, namely carbon, nitrogen,
phosphorus and sulphur, and their cycling in the
estuarine system.

2.3 GEOCHEMICAL CYCLES

2.3.1 Carbon cycle

The major pathways of the carbon cycle are shown
in Figure 2.8. There are two inputs of carbon into
the system, one through allochthonous inputs such as
river flow, runoff and of course waste (e.g. sewage)
disposal, and the second via the photosynthetic
activivies of the primary producers. The carbon

42

Figure 2.8 The carbon cycle in an estuary. R = respiration, P = photosynthesis, C = consumption, E = excretion, M = mortality and D = degradation.

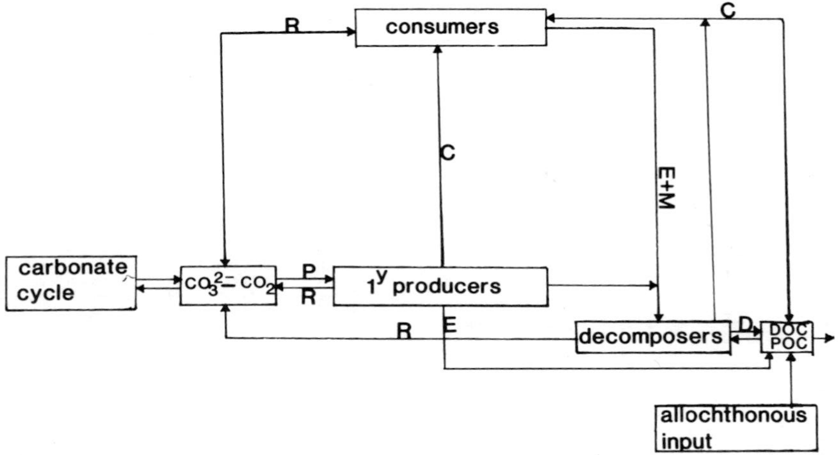

dioxide available for photosynthesis is in
equilibrium with the ions in the carbonate cycle
$$CO_2 \iff H_2CO_3 \iff HCO_3^- \iff CO_3^{2-}$$
with an increase in pH to more alkaline conditions
moving the equilibrium to the right. The extreme
range of seawater pH values are from around 7.6 to
8.3, and at its normal pH of 8.0 to 8.3, the main
ion is bicarbonate (HCO_3^-). In estuaries these
limits may be altered according to the amount and
status of the freshwater inflow, but the presence of
weak acids such as carbonic acid (H_2CO_3) make
seawater a good buffer.

The contribution of the different inputs varies
according to the situation. Biggs (1970) has
calculated that allochthonous input to the Upper
Chesapeake Bay accounts for some 90% of the organic
input into the system. Further down, in the Middle
Bay, the production within the system has increased
dramatically to provide 50% of sytem input, and this
aspect of the primary productivity of estuarine
systems will be considered later.

Allochthonous input arrives as a mixture of
Particulate Organic Matter (POM) and Dissolved
Organic Matter (DOM). As its names suggest, POM is
present in particulate form and includes
phytoplankton and bacteria, but its main constituent
is what is termed "detritus", an all-embracing word
which includes a variety of substances and
microorganisms usually in association with dead and
decomposing organic material. The distinction
between POM and DOM is rather an artificial one and
in practical terms rests on the size of filter used
(commonly 0.5μm or 0.22μm): what passes through the
filter is the DOM.

The amount of Particulate Organic Carbon (POC)
in POM and Dissolved Organic Carbon (DOC) in DOM
depends on the substances involved. Typical values
for river input would be 10 - 20 mg/l DOC and 5 - 10
mg/l POC, which is diluted down in the estuary to
around 5mg/l and 2mg/l respectively. In the open
sea, away from the terrestrial influence, values
for DOC and POC are typically around 1mg/l and
0.1mg/l respectively, and a sewage input into the
system will bring with it a load of around 100mg/l
DOC and 200mg/l POC!

Within the system, there is some leakage of DOM
directly from organisms, but the major cycling
occurs with the decomposition of dead organisms and
faeces and other excretory products (Figure 2.8).
The rate of breakdown of this material depends both
on external factors such as temperature, and its

composition, where the surface area available for microbial action is particularly important, and in this regard both physical breakdown and animal activity especially comminution in feeding play considerable roles.

Godshalk and Wetzel (1978) have produced a generalised equation for the rate of decomposition, k, such that

$$k = (T . O . N)/(R . S)$$

where T = temperature, O = oxygen, N = mineral nutrients available for microbial growth, R = refractility, that is the susceptibilty of the material to breakdown, and S = size (area/volume). They identified three main phases of breakdown:

i) after an initial lag phase, there was a rapid release and metabolism of DOM resulting in a high decomposition rate:

ii) the rate gradually decreased as less and less refractile POM was left:

iii) finally decomposition to all intents and purposes ceased, and the remainder of the material was permanently incorporated into the sediment.

The presence of oxygen is critical to the rate, as aerobic breakdown is more efficient that anerobic. Mechelas (1974) suggested that the decomposer bacteria utilise firstly oxygen and then nitrate, sulphate and finally carbon dioxide in decreasing order of preference as electron acceptors. Oxygen used for breakdown or respiration must be replaced by diffusion from the atmosphere and down through the water column (although there may also be a photosynthetic input where and when there is sufficient illumination). Within the sediment, the depth at which oxygen demand exceeds supply is determined by the amount of microbial activity and the permeability of the sediment. Where there is a great deal of microbial activity, for example in sediments with high organic content which may or may not be due to artificial enrichment as with sewage, the aerobic layer can be limited to a millimetre or so of the surface, or indeed absent altogether in totally anoxic sediments.

The permeability of the sediment is determined by the sizes and sorting of the particles of which it is composed. Sediments with large well-sorted particles with lots of pore space between the grains are highly permeable. In addition, these kinds of sediments are found in areas of high water energy (see Figure 2.6), and both wave and tidal movement help oxygenate the water and to pump it up and down through the sediment. In this situation, typical of sandy shores, the aerobic layer, recognisable by its yellow or brown colour, may reach down some tens of centimetres before changing into the black of the anoxic layer. In contrast, the aerobic layer of estuarine muds may be only millimetres thick.

Corresponding to the change from aerobic to anaerobic conditions is a rapid change in the redox potential (Eh) of the sediment, from about +400mV in the oxidised layer to -200mV in the anoxic zone, and this area is known as the redox potential discontinuity (RPD) layer (Fenchel and Reidl, 1970).

Once the oxygen is exhausted, the organic matter can only be decomposed by the use of nitrate, sulphate or carbon dioxide or bicarbonate ions, which are progressively reduced in that order. Nitrate is used preferentially to sulphate as an electron acceptor because the process yields more energy, although nitrate reduction does not seem to be a major contributor to the overall breakdown budget. In the Limfjord in Denmark, Fenchel and Blackburn (1979) calculated that the more efficient aerobic processes accounted for 75% of carbon oxidation in decomposition, sulphate bacteria for 23% and that a mere 2% was due to nitrate-respiring bacteria.

Sulphate is the second most common cation in seawater after chloride, so it is perhaps not surprising that sulphate reduction is the main anaerobic decomposition pathway. Methanogenesis, which is an important freshwater anaerobic pathway, plays little part in estuarine decomposition. The reduction of carbon dioxide to methane occurs only where sulphate availability is limited, and it appears that the methanogenic bacteria are in fact inhibited by the sulphate bacteria due to competition for the limited amount of hydrogen in saltmarsh sediments (Abram and Nedwell, 1978). In saltmarshes, sulphate reduction can be in the order of 100 g S /m^2/y (Nedwell and Abram, 1978) compared to methanogenesis of 1g CH-C/m^2/y (Atkinson and Hall, 1976).

Cammen (1976) has suggested a carbon turnover time of three to four years for a North Carolina saltmarsh, although the actual rate at which carbon can be re-cycled through the system obviously depends on many factors, not least those limiting the decomposition rate (see equation of Godshalk and Wetzel (1978) above). In temperate latitudes the half-life for the decomposition of material would appear to be anything from days to months, depending on its refractility, but Cruz and Gabriel (1974)

have suggested an average breakdown of 40% per year
for *Juncus* leaves.

The decomposition process also results in the
release of nutrients and other elements which may
then be re-cycled within the system, and of these
two of the most important are nitrogen and
phosphorus because of their role in primary
production.

2.3.2 Nitrogen cycle

The nitrogen cycle (Figure 2.9) has the same
basic relationships between the consumers, primary
producers and decomposers as noted above, but there
is an important difference in that the availability
of nitrogen is the limiting factor for much of the
estuarine primary production, particularly in
saltmarshes (Valiela and Teal, 1974). From the
decomposers, nitrogen is released as ammonia and
transformed first to nitrite by the bacteria
Nitrosomonas and then to nitrate by *Nitrobacter*.
Nitrification is an oxygen dependent (aerobic)
process, and consequently there is little
nitrification in anaerobic sediments. In the absence
of oxygen, nitrate is denitrified and reduced to
atmospheric nitrogen. In this form, the nitrogen
must either be fixed by nitrogen-fixing bacteria or
it is lost to the system by diffusion. The process
can work the other way as well, and the system can
gain nitrogen from the atmosphere by such fixing
(Figure 2.10). Jones (1974) showed N fixation by
blue-green algae to be up to 40 g N/m²/y in an
English salt-marsh, and rates of 1 g N/m²/y to 11.5g
N/m²/y have been found in American and Canadian
marshes (Whitney et al., 1975; Patriquin, 1978).
Some of the highest rates of N fixation, some 180
mg/m²/d equivalent to over 60 g N/m²/y have been
reported from coral reefs (Webb et al., 1975).

The nitrogen budget of Great Sippewisset Marsh
in Massachusetts (Valiela and Teal, 1979), shows
however that nitrogen-fixing was not the main input
into the sytem, and in fact the amount fixed was
under half that lost through denitrification (Figure
2.10). Allochthonous inputs from groundwater and to
a lesser extent precipitation were also
considerable, but the major movements of nitrogen in
and out of the system were tidally mediated, and
overall the system was exporting nitrogen to the
surrounding waters. There was also a loss from the
system to the sediment, although this could

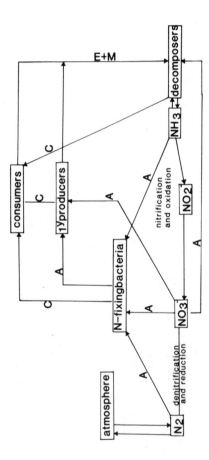

Figure 2.9   The nitrogen cycle in an estuary. A = assimilation, other notation as Figure 2.8.

presumably become available again in time. The sediment also has a most important role in providing this pool of nitrogen for the system, and it should be noted that the size of the sediment nitrogen pool is almost an order of magnitude greater than the annual inputs.

Figure 2.10 Nitrogen budget ($kg/km^2/y$) for Great Sippewisset Marsh, Massachusetts, USA (from Valiela and Teal, 1979). Sediment values include dead organic matter and detritus; primary producer biomass is given as the maximum (August) figure.

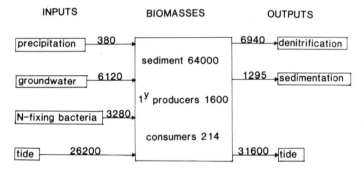

In terms of nitrates for plant growth, Henriksen and Jensen (1979) found in a Danish saltmarsh that it was almost all generated within the sediment and that there was very little tidal contribution: rather there was a net export as in the American example.

The importance of the sediment to the nitrogen budget makes it difficult to be more specific in general terms, as the cycling and fluxes will depend largely on local conditions such as the degree of aeration of the sediment. Even within the one location there will be considerable variability with time, notably with seasonal cycles of input and loss, and with unpredictable events such as the amount of material swept away in storms.

2.3.3 Phosphorus cycle

The comments on nitrogen apply equally to the phosphorus cycling and budgets (Figures 2.11 and

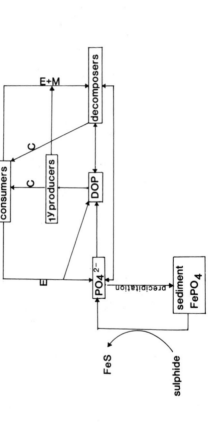

Figure 2.11 The phosphorus cycle in an estuary. Notation as Figures 2.8 and 2.9.

2.12 respectively). The phosphorus cycle is simpler than the nitrogen cycle with phosphate as the the form in which it is taken up by organisms and also the form in which it is liberated by excretion or through the deocomposition of organic material. The availability of phosphorus is frequently a limiting factor in primary production, and as a consequence plants have developed extremely efficient mechanisms of phosphorus uptake. As a result, phosphorus that remains within the system may be cycled extremely rapidly, but phosphorus may be lost to the system or made temporarily unavailable through the formation of organic esters or precipitation into the sediment as $FePO_4$. The subsequent re-availability of these forms depends on the degree of microbial activity in the sediment (Fenchel and Blackburn, 1979) and it should be noted that the mediating process for one of the liberating mechanisms, that of sulphide to FeS, occurs under anaerobic conditions.

Figure 2.12   Phosphorus budget (kg/y) of a Mexican lagoon (from Arenas and de la Lanza, 1983).

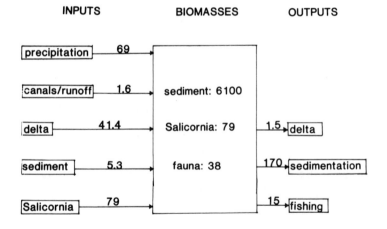

The example of a phosphorus budget (Figure 2.12), underlines the part played by the sediment as a reservoir of phosphorus, and as the major loss to the system. Note also that the annual input from *Salicornia* is in fact its total production - as in Figure 2.10, the biomass figure quoted is the maximum standing stock attained. The example is a little different from most systems studied in that there is a net gain from marine sources, in this case the delta. In other situations such as a temperate saltmarsh (Woodwell et al., 1979), there was a net loss of phosphorus, and Reimold (1972) considered that the *Spartina* acted as a nutrient pump, transferring phosphorus from the sediment to the surrounding waters. Using labelled phosphorus, Reimold (1972) found that this transfer rate was at its peak some 10 to 15 days after the introduction of the isotope, giving a maximum loss of 6 kg P/ha/d at the peak growing season, and that almost all was lost from the sediments within a month. It has been suggested (Odum, 1980) that the status of an estuary as an importer or an exporter, and indeed the degree to which these terms apply, is related to the tidal exposure: macro-tidal estuaries are exporters while micro-tidal estuaries are importers.

There are two important points to emphasize in this brief outline of nutrient cycling. The first point is that the limiting factor for a particular step such as primary production is not the amount in the system, but the amount available at any one time. This distinction is made clear by a consideration of the sediment nitrogen and phosphorus pools (Figures 2.10 and 2.12) compared to the amount directed through the primary producers. Patriquin (1972) compared the amounts of available N and P in the sediments with the daily requirements of the *Thalassia* bed and concluded that while there was over a year's supply of P, there were only a few day's supply of N, leading to the conclusion that in this situation, N was more likely to be the limiting factor.

The second point is the part played by the sediment and the importance of oxygenation and the redox status. When a sediment becomes anaerobic, the decomposition process slows down, nitrate production via nitrification ceases and existing nitrate is denitrified to gaseous nitrogen giving a double loss, and precipitated $FePO_4$ is liberated as $PO_4{}^{2-}$. The factors determining the oxygenation of the sediment have been outlined earlier, but the importance of biological activity should also be mentioned. Those plants which have root systems, such as *Spartina*, *Salicornia* or *Zostera* in temperate systems or mangroves in the tropics, will transport oxygen into the sediments by diffusion out of the root tissues. The importance of the aerobic zones around the roots has been emphasised by Patriquin (1978) who estimated that while the rate of N

fixation at the mud surface was 2.2 $g/m^2/y$, the rate
in the *Spartina* rhizosphere was 9.3 g $N/m^2/y$.

Photosynthesis will also raise the oxygen
content of the water (supersaturation) overlying the
sediment, but it must be appreciated that this is a
two-way process; what is gained by extra photosynthesis during the day may be consumed by extra
respiration by the same organisms during darkness.
Likewise extra biological or chemical oxygen demand
(BOD or COD) as a result of waste input will
increase the tendency toward sediment anoxia.

The effect of animal activity may also be
considerable. Physical turnover of the sediment,
either during burrowing or feeding, including
comminution or other changes in the passage through
the gut brings anoxic sediment to the surface and
promotes microbial activity and breakdown of organic
matter. Animals which live in the sediment aerate
the deeper layers, partially by oxygen diffusion out
of their tissues, but largely through the
construction of burrows and their irrigation with
water, such that close examination of even the
deepest burrows reveals a light brown oxygenated
sediment layer lining it all the way down (Anderson
and Meadows, 1978).

However, depending on a variety of factors such
as water depth, sediment type and the quantity of
other (allochthonous) inputs, there is some debate
as to the amount of the contribution actually made
in this way to the nutrient budget of the system.
Rowe et al. (1975) concluded that benthic
regeneration was a major factor influencing coastal
primary production, but Hartwig (1976) concluded
that nutrients released from the bottom were
superfluous to the needs of the phytoplankton.

2.3.4 Sulphur cycle

It is only the larger animals, the macrofauna,
which construct burrows. The smaller animals,
termed meiofauna, those too small to be retained by
a 1mm seive, live in the interstices between the
sediment particles and are therefore subject to the
oxgygen regime according to the depth at which they
live. The majority of meiofauna are to be found in
the aerobic layers, and are particicularly abundant
in the oxidised sediments lining macrofauna burrows,
but some groups are restricted to areas below the
redox potential discontinuity layer (RPD) and are
obligate anaerobes with a sulphur based metabolism.

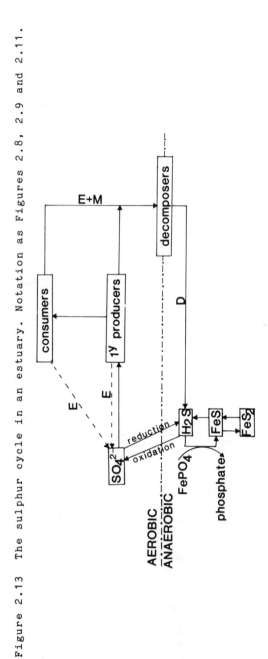

Figure 2.13 The sulphur cycle in an estuary. Notation as Figures 2.8, 2.9 and 2.11.

Such eukaryotes are interesting examples of what may be an evolutionary primitive condition (Boaden, 1977), but the principals in the sulphur cycle (Figure 2.13) are the prokaryotes, the bacteria.

Sulphur present in the sediments as hydrogen sulphide may be oxidised to sulphate by thiobacilli whose twin requirements of oxygen and low oxidation state sulphur restrict them to the vicinity of the RPD layer. Reduction of sulphur and sulphate may be carried out by obligate anaerobes such as *Desulphovibrio*, along with the formation in anaerobic conditions of a variety of iron-sulphur compounds. Their formation is accompanied by the reduction of ferrous phosphate and the liberation of phosphate (see also Figure 2.11).

In addition to the link with the phosphate cycle, the sulphur cycle is also closely connected to the carbon cycle and the decomposition process, as already mentioned, as well as the nitrogen cycle again with reference to decomposition and the part many sulphur-reducing bacteria play in nitrogen fixation. Sulphide is also a strong ligand, and metal ions that have been bound into largely insoluble metal sulphides tend to be locked into anaerobic sediments. In polluted situations, the sulphide cycle thus assumes an extra importance in the cycling and scavenging of heavy metals.

2.4 PRIMARY PRODUCTION

So far the discussion of the biological components of the estuarine system has been centred on the part played by the microbes, and in particular by the part they play in making substances available to the higher organisms and in the first trophic step in the system, primary productivity.

Primary producers come in an immense variety of shapes and forms, from microscopic single-celled algae and diatoms on mud surfaces to mangrove trees many metres in height. Table 2.1 summarises some of the major types of primary producer, and gives rates of production for each. It is interesting to note that despite the variety of types and locations the range of values is relatively restricted and indeed there is almost as much variation within one type (*Spartina*) as there is between types.

The ultimate source of energy to the primary producers and through them to the system is from sunlight, but it can be seen from Table 2.1 that

Table 2.1 Primary productivity (kJ/m^2/y) in estuaries and coastal waters

Community	Species	Location	Productivity	Reference
Saltmarsh	Salicornia	UK	18900	Jeffries (1972)
	Spartina	UK	17300	Ranwell (1961)
	Spartina	Georgia, USA	68400	Odum & Fanning (1973)
	Spartina	N.Carolina, USA	11700	Jeffries (1972)
Sea-grass	Zostera	Denmark	19700	Sand-Jensen (1975)
	Zostera	N.Carolina, USA	11900	Thayer et al. (1975)
	Posidonia	Mediterranean	43400	Drew (1971)
	Thalassia	Florida, USA	30600	Jones (1963)
Mangrove	Rhizophora	Florida, USA	11900	Heald (1969)
Macroalgae	Macrocystis	California, USA	11900 - 51000	Wheeler (1978)
	Laminaria	UK	41700	Bellamy et al. (1968)
	Fucus	Massachussetts, USA	153000	Kanwisher (1966)
	Enteromorpha	UK	18400	Baird & Milne (1981)
Microalgae	--	UK	4900	Joint (1978)
	--	Louisiana, USA	8200	Wolff (1977)
Phytoplankton	--	UK	2800	Joint (1978)
	--	Netherlands	4400	Wolff (1977)
	--	Louisiana, USA	7100	Wolff (1977)
	--	continental shelf	6200	Mann (1982)
Coral reef	various	Florida, USA	33500 - 126600	Kanwisher & Wainwright (1967)
Terrestrial	grassland	temperate	3600 - 26700	Krebs (1978)
	rainforest	tropical	17800 - 62300	Krebs (1978)

northern systems (e.g *Zostera*, Denmark) are almost
as productive as southern (*Thalassia*, Florida).
Since Florida receives much more light energy than
Denmark, there must be some other factor limiting
production and this can be best explained by
reference to Figure 2.14.

Figure 2.14a shows the conventional cycle of
phytoplankton production in the open sea.
Productivity is low at the start of the year, but
increases as the availability of light increases in
the spring. As the phyoplankton increase, the
nutrients become locked into the plant tissue, and
so the nutrient levels fall until they become the
limiting factor for production. This process is
exacerbated by the stratification of the water
column since the thermocline effectively restricts
nutrient cycling to the top few metres. Following
the spring peak, production continues at a low level
throughout the summer until more nutrients become
available through a die-off in the phytoplankton or
a mixing of the water by autumn gales which bring up
the nutrient-rich bottom waters. This secondary
bloom, which is usually much smaller than the spring
bloom, continues until the falling light levels
again become limiting.

The situation in most estuaries and some
nearshore coastal environments (Figure 2.14b) is
somewhat different (Schlieper, 1971). Here light
levels may be severely restricted because of the
turbidity of the water, but there is a plentiful
supply of nutrients from a variety of sources
including the sediment. Consequently, production
(which is mainly benthic) is not necessarily
nutrient limited, (but see comments in Section 2.3.2
on N relations and saltmarshes) and so productivity
continues, and continues to increase throughout the
summer. It is worth noting however that the overall
primary productivity may be partitioned between
different species or indeed different types; in the
Bull Island lagoon system for example there is an
initial production of green algae followed somewhat
later by the *Salicornia* (Madden, 1984). As a result,
the shape of the primary productivity curve follows
that of light intensity fairly closely.

In tropical waters, the situation is slightly
different, in that there is an almost constant
year-round supply of light. In this situation, the
nutrients can get locked into the biota, and
production becomes more or less constant throughout
the year. Nevertheless, productivity in the tropics
can be high, as shown by the data for mangroves and

corals in Table 2.1.

In coral systems, although these are not truly estuarine, they avoid loss of nutrients from their system by cycling them through the coral/ zooxanthellae symbiosis, and although the gross primary production may be high, the respiration of the community is also high, indicating that much of what is produced is utilised within the community.

Figure   2.14   Seasonal   variation (spring, summer, autumn, winter (SSAW)) in primary productivity (solid line), light (dotted line) and nutrients (dashed line) (arbitrary units) in a) open sea and b) estuarine conditions (temperate waters).

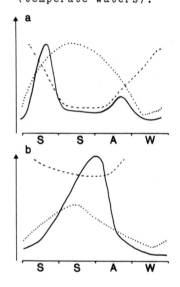

2.5 SECONDARY PRODUCTION

The corals, because of this coral/zooxanthellae symbiosis, are rather a special case however, and in

Table 2.2 Secondary productivity (kJ/m²/y) in the estuarine and coastal system

Feeding type	Species	Location	Productivity	Reference
Deposit	meiofauna	UK	927.8	Warwick et al. (1979)
	Macoma	UK	179.8	Chambers & Milne (1975b)
	Macoma	Nova Scotia, Canada	34.4	Burke & Mann (1974)
	Macoma	California, USA	1104.4	Nichols (1977)
	Scrobicularia	UK	520.8	Hughes (1970)
	Hydrobia	Netherlands	93.9	Wolff & de Wolff (1977)
	Littorina	Georgia, USA	170.5	Odum & Smalley (1959)
	Arenicola	Netherlands	177.7	Wolff & de Wolff (1977)
	Palaemonetes	Massachussetts, USA	306.6	Welsh (1975)
Filter	Modiolus	Georgia, USA	70.1	Kuenzler (1961)
	Mytilus	UK	4770.4	Milne & Dunnet (1972)
	Mytilus	Nova Scotia, Canada	350.7	Burke & Mann (1974)
	Mya	Nova Scotia, Canada	206.7	Burke & Mann (1974)
	Mya	California, USA	357.9	Nicholls (1977)
	Cerastoderma	UK	1692.2	Hibbert (1976)
	Gemma	California, USA	1094.9	Nicholls (1977)
	Crassostrea	S.Carolina, USA	7191.2	Dame (1976)
Grazer	Littorina	Nova scotia, Canada	26.7	Burke & Mann (1974)
	Orchilimum	Georgia, USA	43.5	Odum & Smalley (1959)
	zooplankton	North Sea	735.0	Steele (1974)
Predator	Nephthys	UK	130.7	Warwick & Price (1975)
	Carcinus	UK	32.2	Baird & Milne (1981)
	Platichthyes	UK	23.0	Baird & Milne (1981)
Omnivore	Nereis	Belgium	1142.8	Heip & Herman (1979)
	Nereis	UK	229.1	Chambers & Milne (1975a)
	Fundulus	Delaware, USA	724.5	Meredith & Lotrich (1979)
Terrestrial	rabbit	UK	740.0	Spedding (1977)

most sytems, the transfer of energy is to the
consumer trophic level. A selective list with
productivity values is given in Table 2.2. The
majority of the examples given are deposit or
detritus feeders, reflecting the mode of energy
transfer up through the system. Kirkman and Reid
(1979) quantified the fate of the primary production
of a *Posidonia* bed in Australia, and found that 37%
was consumed by benthic detritivores and only 3% was
actually taken by the grazers. Of the rest, some 12%
was lost as POM and an astonishing 48% as DOM. Odum
and Smalley (1959) likewise found that the energy
flow in a Georgia salt-marsh through the detritus
feeding *Littorina* was an order of magnitude greater
that that of the grazer, the grasshopper *Orchilimum*.
An exception can probably be made with seaweed based
sytems on rocky substrates in fjordic systems for
example, where herbivores can affect both the
productivity and the structure of the community
(Paine, 1977).

In fact it is difficult to differentiate clearly
between the feeding categories, and detritus finds
its way into the diets of most estuarine species
(see for example the various contributions in Jones
and Wolff (1981)). The polychaete *Nereis
diversicolor* is a typical example. It has the large
strong jaws of a predator, but it also filter feeds
from the water current through its burrow as well as
taking in quantities of sediment, although it is
unclear whether this is directly eaten or more
probably filtered out and eaten with other water
borne particles.

Similarly many other filter feeders will take
in sediment if it happens to be in suspension, and
likewise the deposit feeders such as the bivalves
can collect their food by filtration. The snails
such as *Hydrobia* feed by grazing the mud surface,
and take in whatever mixture of microalgae, diatoms,
detritus, sediment and bacteria happen to be
present. The system is not totally haphazard,
however, and different species of *Hydrobia* for
example, can show sophisticated selection mechanisms
for their preferred substrates to reduce competition
for this resource (Barnes, 1979). Finally many
species can take up DOM such as amino acids against
the concentration gradient, so that even the DOM
from the primary producers could be directly
utilised, but it is difficult to quantify this in
terms of the overall budget.

2.6 ENERGY FLOW

Because of the difficulties referred to above, the examples of food webs, one from a European estuary dominated by mud-flats, the Ythan estuary in Scotland, and one from an American estuary at Beaufort, North Carolina dominated by sea-grass, shown in Figure 2.15 are gross simplifications, and can only identify what are considered to be the major elements and pathways.

Nevertheless, there is a great deal of similarity between the two systems, the more so since the organisms have been grouped according to function. Although the actual species, and in some cases even the phyla, may differ from the one to the other, they are carrying out the same duties and so the pathways and transfers and the overall structure are similar.

The major difference is in the primary productivity, in that the European example is dominated by phytobenthic production (of which the green alga *Enteromorpha* accounted for more than 60%), and the American by the seagrass, with the algae playing a relatively minor role. Note that the contribution of phytoplankton varies, from about 30% in the seagrass dominated system to an insignificant 2% or so in the Ythan. The values shown for phytoplankton productivity in Table 2.1 bear out this lesser role in comparison with other types and reinforce the concept of estuaries as sediment rather than water column based systems.

In both systems very little of the primary production goes directly to the grazers, the herbivores in the system, and most is transferred to the higher trophic levels via the decomposers. In this the link between the transfer of energy through the system and the cycles previously described (notably C, N and P) is obvious. The main difference between energy flow and nutrient cycling is that the major source of energy (excluding the relatively minor role played by chemotrophs) is sunlight, and there is also a loss of energy to the system through respiration (for clarity this has been omitted from Figure 2.15). Consequently one talks of nutrient cycling but energy flow through the system.

Odum (1980) considered that the hypothesis that estuarine and salt-marsh systems were detritus rather that grazing food chain based was sufficiently well established and proven from a variety of situations to be acceptable as an "emergent property", that is one which transcends

Figure 2.15 Energy flow (kJ/m^2) through two estuarine food webs: the Ythan, Scotland (left) and the Newport, Beaufort, USA (right).

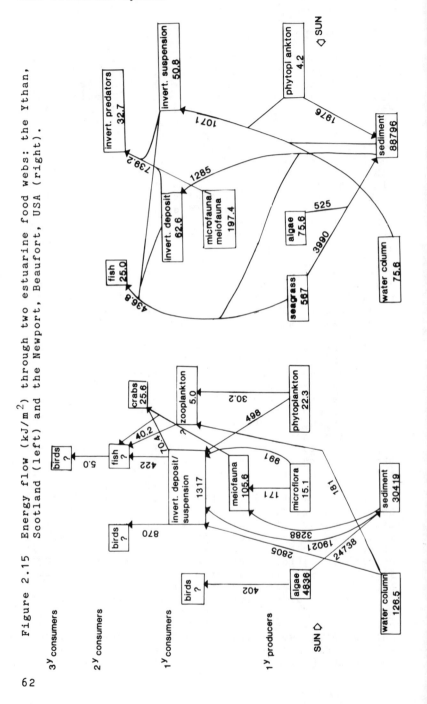

mere observation of the system components, although he did caution that the detritus complex itself comprised a poorly-understood and little-studied microcosm of autotrophs and heterotrophs.

The efficiency of the transfers between the different trophic levels varies greatly. Photosynthetic efficiency, the efficiency with which the plants fix the energy from the sun is typically around 2%, and as a rough rule of thumb, ecologists usually assume that about 10% of the energy from one level is transferred to the next (Meadows and Campbell, 1978). However it is apparent that in many cases, the efficiency is much greater than that, and Baird and Milne (1981) have calculated that the efficiency of transfer between the suspension feeding macroinvertebrates and the birds in the Ythan exceeds 70% (see Figure 2.15). This is clearly an upper figure, and in the other example in Figure 2.15, Thayer et al (1975) calculated that the fish only consumed about 7% of the annual net primary production. Transfer efficiency from one level to the next may well be greater on average than previously thought, perhaps around 20% - 25% (Krebs, 1978).

The efficiency with which energy is transferred up through the system depends on both the ability to obtain the food, the efficiency with which it is digested or assimilated and the efficiency with which this is converted to body tissue i.e. production. Figures for a variety of mollusc species have been drawn together by Davis and Wilson (1985) which suggest a mean value for net growth efficiency of around 30% (Table 2.3). It is important to realise that the unassimilated portion of the food, the faeces, is rarely lost to the system, but is recycled via the detrital decomposers (see the carbon cycle, Figure 2.8).

Energy is lost to the system through the heat output of respiration. Losses due to respiration vary also, but a general figure for invertebrates equivalent to some 50 - 90% of assimilation has been suggested (Table 2.3).

In the examples given above (Figure 2.15), invertebrate respiration in the Ythan and at Beaufort totalled some 8620 kj/m^2/year and 2833 kj/m^2/year (including microbiota) respectively. Far higher losses are incurred by warm-blooded animals due to the costs of maintaining body temperature, and the energetic costs of flight in birds increases this significantly. For this reason many birds lose condition over the winter despite the apparent abundance of food, and disturbance, as we shall see in a later chapter, can seriously affect their chances of survival.

Table 2.3   Net growth efficiency (NGE = Total Production x 100/Assimilation) and respiration assimilation (RA = Respiration x 100/Assimilation) of selected marine molluscs (Davis and Wilson, 1985)

Species	NGE	RA
T. funebris	14	86
P. vulgata	25	75
L. irrorata	14	86
L. littoralis	40	68
S. plana (mean)	23	78
N. turgida	36	64
T. fabula	43	57
M. mercenaria	36	64
C. islandica	39	61
M. demissus	32	68
Mean	30.2	70.7

Also omitted from Figure 2.15 are the not inconsiderable inputs (in the form of DOM and POM) from river and sea, and to this can often be added waste discharges. In the Ythan example above (Figure 2.15) river and sea input exceeded primary production several times over, although this probably gives a false picture, as the sea took away as much if not more with the outgoing tide as it had brought in. Nevertheless this water-borne POM remains a potential food source for many estuarine organisms.

At this stage it may be useful to offer a conceptual summary of the estuarine system as a "black-box" system as depicted in Figure 2.16. Matter (or energy or nutrients) enters it and is stored, then often changed physically, and subsequently re-released to the sea or the atmosphere. This concept is particularly useful in terms of management if we imagine that the estuary acts as a sink, for sedimentation (Figures 2.6, 2.7), nutrients (Figures 2.9 - 2.12) and energy (Figure 2.15), such that any change in the system in terms of input or output is buffered both temporally and to a lesser extent spatially. For example, the reserves of phosphorus in the sediment (Patriquin,

1972) could if mobilised keep production going for
some years even if all P inputs (including natural
inputs!) were stopped immediately.

Figure 2.16   The estuarine system as a "black box".

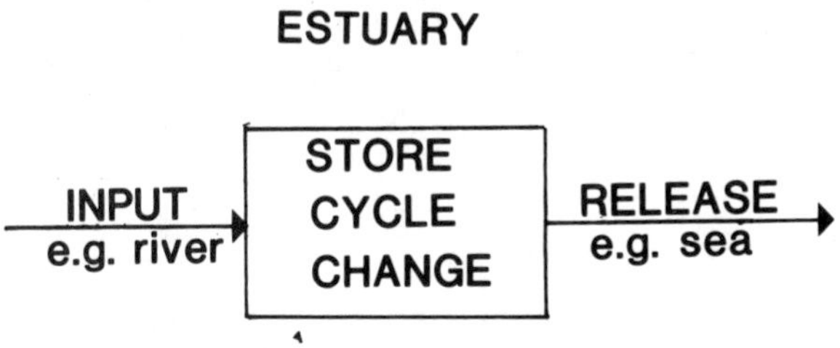

ESTUARY

Despite these often considerable inputs, more
leaves with the outgoing tide than comes in with the
flood and productive estuarine systems export
organic matter to the surrounding offshore waters.
Odum (1980) considered this as the second of the
ecosystem-level hypotheses to attain the status of a
collective property, and used the term "outwelling"
to describe it. There are situations where
outwelling is weak or undetectable, but Odum (1980)
considered that these principally arose where
communication between estuary and sea was restricted
- for example by narrow, shallow sills typical of
fjordic type estuaries - or where tidal action is
weak (see for example the P budget in Figure 2.12).
The narrow enriched zone of the coastal front has
been described extending some 10 km offshore on the
coasts of Georgia and Louisiana, USA where the
estuaries are typically bar estuaries (Figure 2.1b),
and the effect may be even more pronounced in
coastal plain estuaries (Figure 2.1a) where there is
unrestricted exchange between the estuary and the
sea (Odum, 1980).
Nevertheless it must be emphasised that even in
those estuaries in which outwelling is considerable,
only a small fraction of the total budget is
exported, and the great majority is recycled round
the system. In this it is not so different from the
corals which were referred to earlier, except that
in the coral system the recycling is much more
restricted (in terms of the number of transfers) and
consequently comparatively less is lost.

The estuarine system

The third of Odum's (1980) ecosystem-level hypotheses was the tidal subsidy hypotheses, that is a significant correlation between primary production of saltmarshes and tidal amplitude or range; the greater the amplitude, the greater the production, at least within a moderate range of amplitudes. Odum (1980) considered that tidal action removed wastes and increased nutrient fluxes and cycling, all of which was to the benefit of the organisms enabling them to devote more solar energy through photosynthesis to production.

2.7 ENVIRONMENTAL STRESSES

This last proposition is perhaps the most surprising, as the tide is one of the many stresses to which estuarine organisms are subject, and Odum (1980) indeed first outlined it as a subsidy-stress syndrome and noted that extreme tides, for example in the Bay of Fundy were not likely to result in enhanced production.

One principal consequence of the stresses of the estuarine system is the restricted number of species found in estuaries. A profile down an estuary from river to sea results in the kind of species distribution shown in Figure 2.17 (after Remane, 1971). River or truly freshwater species are restricted to salinities of less than about 0.5 ‰ and likewise marine species penetrate no further upstream than the 30 ‰ isohaline. In between there are the euryhaline brackish water species, relatively few in number but often present in great abundance or biomass. The main stressor is salinity, but it is important to realise that there are others, notably those associated also with life in the littoral zone, namely temperature, light (restricted by turbidity) and desiccation through tidal exposure. To these are added such factors as oxygen deficiency, which may be exacerbated, as will be seen in later chapters, by waste inputs.

It is also important to realise that it is the fluctuations of these environmental factors and the rapidity of the fluctuations which impose the stress on the organisms. To take the example of the salinity, species such as *Macoma balthica* are found in the Baltic at salinities below those which limit their distribution in estuaries, because the salinity of their Baltic habitat is not subject to the same degree of diurnal or seasonal change.

The majority of work on estuarine organisms

deals with the benthic flora and fauna, for reasons which have already been mentioned. Organisms of the estuarine water column, the plankton (and nekton) can avoid some of the worst of fluctuations as they remain in the same water body which, although it may move in and out with the tide, remains relatively constant in composition. Only the larger estuaries tend to have more than a transient planktonic community, and even this, for example the copepod *Eurytemora affinis*, may show an affinity for sediment in the form of the turbidity maximum (Soltanpour-Gargari and Wellershaus, 1985).

Figure 2.17 Distributions of freshwater (FW), brackish and marine (SW) species with distance down the estuary (after Remane, 1971).

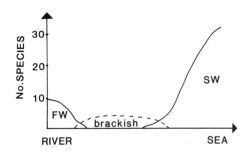

Where the external environment is one which changes, the organism has two strategies open to it: it can allow its internal state to change to match that outside or it can maintain its internal state. These two strategies are called "conformer" and "regulator", and are depicted in Figure 2.18.

There are advantages and disadvantages to both strategies. The advantage of maintaining the internal body state constant is that conditions for metabolism and so on can be maintained nearer to their optimum - for example the enzymes in warm-blooded creatures work best at $37^{\circ}C$ which is the body temperature they try to maintain. The disadvantage is that maintenance of body temperature demands a high energetic cost, as has been mentioned already in estuarine birds, and there may be situations at either end of the temperature scale where the homeostatic system cannot cope and cannot maintain temperature.

Figure 2.18   a) - regulation and conformity; b) -
example (axes as a) showing osmo-
regulator (*N. diversicolor*) and osmo-
conformer (*A. marina*): note convergence
at 35‰ salinity (full seawater).

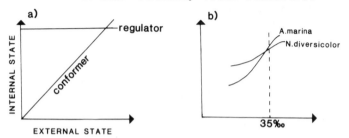

The same situation exists in respect of response
to other variables, and Figure 2.18 shows the
response of the polychaete *Nereis diversicolor* to
changes in salinity. The point to note is that
although *Nereis* does regulate, regulation is not
100%, and the range of salinities over which it
regulates matches closely to those in which it is
found in nature. Outside this range, Nereis either
does not or cannot regulate, and effectively behaves
as a comformer. Some fish, of which the salmon is
probably the most notable example, can regulate the
internal/external exchange to allow the transition
from salt to fresh water and vice versa, but they
are relatively few in number, and in others such as
the eel, the change is one-way and irreversible.
The process of osmoregulation and the mechanisms by
which it is achieved have been studied in
considerable depth over the years, and here again
the reader is directed to the many texts on the
subject (e.g. Rankin and Davenport, 1981).

An example of the alternative strategy is shown
by another polychaete, *Arenicola marina* (Figure
2.18). Here the internal state follows that outside,
although again the strategy is not 100%.

There are various ways in which organisms can in
a sense protect themselves from environmental
stresses, and these include both behavioural and
physiological mechanisms. Many bivalves such as
cockles or mussels for example can close their
shells, a response which stops them drying out at
low tide, but it can also be used to exclude harmful
waters, including not only those of low salinity but
also those dosed with contaminants (Wilson 1983).
Physiological mechanisms include the production of

68

thick coats of mucous, to slow down the rate of
equilibration between inside and out, and other
mechanisms such as burrowing in the sediment employ
this principle of somehow insulating the organism
from the worst of the external stresses. This is
an important point: regulation of a substance by
whatever mechanism will obscure and confuse the link
between the amount in the system and the amount that
actually reaches the tissues of the organism and
will thus make very difficult the prediction of
cause/effect relationships of contaminant inputs.

Davenport (1985) put forward the interesting
theory that these mechanisms of regulation, both
physiological and behavioural, which may protect
against a variety of stresses, originated as a
defence against predators, noting in evidence that
estuarine prey species (e.g. bivalves) tend to be
osmoconformers and estuarine predators (crabs, fish)
regulators.

However, it is not only natural stresses that
are imposed in many estuaries, and in the next
chapter we shall examine some more recent stresses
and their effects - the contaminants.

Chapter Three

IMPACTS, CONTAMINATION AND POLLUTION

What is pollution? Strictly speaking it is the impairment of the estuary by harmful substances introduced by man's activities, but often the term is applied to any degradation, real or perceived, to the estuarine environment, and so varies according to the viewpoint or standpoint of the observer.
GESAMP (1986) has defined marine pollution in the following terms:
"Pollution means the introduction by man, directly or indirectly, of substances or energy into the marine environment (including estuaries) resulting in such deleterious effects as harm to living resources, hazards to human health, hindrance to marine activities including fishing, impairment of quality of use of sea water and reduction of amenities".
It follows then that the introduction of a substance into the estuarine environment should properly be termed "contamination" until deleterious effects or damage can be shown: in practice, however, the two terms are often used interchangeably.

3.1 IMPACTS

In the context of this volume, man's impacts on the estuary can be summarised as change of use, over-use and misuse or abuse, with only the final category truly qualifying as pollution as defined above.

3.1.1 Change of use

Change of use covers the topics dealt with in

Chapter One, where the different demands on the estuary have led to changes in the physical form and these in turn have led to hydrodynamic and other changes in the system. In view of the important role played by the sediment in the estuarine system, any changes in the pattern of sediment deposition or sediment character will have wide-reaching effects on the way that the system functions. As an example, after tidal flat or saltmarsh reclamation, river-borne sediment will be deposited elsewhere, and many estuarine ports rely on extensive and expensive dredging operations to remove these unwanted deposits. Changed hydrodynamic conditions and water circulation may also lead to scouring or erosion of banks and coastal margins, and cause problems for navigation and loss of foreshore as well as adding to the sediment load in the water column. Dredging of ports and harbours of this accumulated sediment may also cause damage not only by the physical disturbance involved but also by the re-mobilisation of contaminants which had become incorporated into the sediment.

Dams and barrages likewise change the system. These may be erected for coastal protection against flooding or erosion and can be permanent or moveable for protection against the worst of conditions - storm surges for example. The prime examples of such works are to be found in the Netherlands, where a large proportion of the coastline is protected. The magnitude of the changes to the system depends on the scale of the work; the greater the change from the original state, the more the system is altered. In the case of the Thames Barrage in London, which will only be lowered against severe storm surges, the alteration to the system is minimal, but in the case of the Grevelingen estuary in the Netherlands, which was completely impounded, the change has been from a fairly typical estuary to a saline lake.

The physical, chemical and biological evolution of Lake Grevelingen has been well documented (Neinhuis, 1978), and show how far such alterations go beyond mere hydrodynamic changes or salinity reduction. For example, the major effect on the biological system was that the input of organic matter from the North Sea was cut off, and despite a considerable increase in the phytobenthic production, the amount available to the consumers declined by roughly 40%. Species balance changed too, and these changes were reflected up the food chain, with increases of over an order of magnitude in consumption by both herbivorous and piscivorous

birds.
     The effects of such large scale works can be
lessened by careful study of the options. Leentvaar
and Nijboer (1986) considered the options for
barrages in the eastern Scheldt estuary in the
Netherlands, and the decision from the resultant
policy analysis was a combination of the more
economic scheme with minimal changes to the system
through operating an extended tidal cycle rather
than simply reducing tidal amplitude.
     Coastal protection or reclamation schemes like
these can be found in many industrialised estuaries,
but dams or barrages have also been proposed for
hydroelectric generating. Although the only large
scale commercial tidal hydroelectric station is at
La Rance in France, other pilot schemes have been in
operation, especially in those areas with a large
tidal amplitude such as the Bay of Fundy in North
America. Here again, there will be changes to the
system, and these must be carefully considered to
ensure that present beneficial uses are not
imperilled, or if they are, then the benefits must
outweigh the costs! The collapse of the sardine
fishery at the mouth of the Nile has been linked to
the decline in productivity of the region following
the construction of the dams and the diminution of
flow, particularly floods, on the River Nile,
prompting similar fears, given the exportive status
of estuaries (see Chapter Two), elsewhere.
     Tidal energy is only one of a number of ways in
which energy can be extracted from the marine
environment, and the implications of energy
development in the marine environment has been
reviewed by GESAMP (1984a). Considerable research
has gone into Ocean Thermal Energy Conversion
(OTEC), particularly in developing countries, but
since one of the prime requirements of OTEC is deep
water with a large thermal gradient, the others,
namely biomass conversion and wave energy are more
likely to have a direct impact on estuaries.
     Large scale wave energy developments will affect
the degree of exposure of coastlines in the lee of
such schemes and possibly change the hydrodynamic
and sediment transport and deposition regimes. To
date, only small scale plants have been installed,
with encouraging results as regards the generating
costs, but it is still too early to say whether they
will have any drastic or permanent effect on the
environment.
     The harvesting of algae for biomass energy will
pose the same problems as with any other renewable

resource such as the fishing industry.

3.1.2 Overuse

Overuse is much less prevalent than change of use, and the identification of overuse of estuarine resources is complicated in many cases by the effect of pollution, such that it is difficult to attribute the decline in a resource to over-exploitation alone. An exception can be made for physical resources where, in a few estuaries, sand or gravel may be taken by extractive industries. These resources are often non-renewable, and in addition their removal may disturb the pattern of sedimention along the coastline and accelerate or exacerbate erosion.

The major estuarine resources at risk from overuse are features such as sand dunes. These are fragile and easily damaged environments which are coming under increasing pressure, particularly from leisure activities. They are important habitats in their own right, but again damage here leads to damage elsewhere either from increased erosion behind the former dune barrier, or sand encroachment on lands behind.

The principal biological resources to be exploited are the fish, particularly migratory fish such as the salmon, and the shellfish. The latter are becoming increasingly important through mariculture, but it is doubtful whether they were ever exploited to the point at which stocks were in danger. On the other hand, the salmon catch in many, if not all, estuaries has declined, but it is a moot point as to whether this is due to fishing pressure within the estuary, or a combination of over-fishing at sea, destruction of spawning grounds upriver and pollution everywhere (Clark, 1987).

Apart from commercial fishing, sport fishing can have serious effects particularly where the pressure on the resource is intense. Damage by sport fishermen includes the damage done by bait digging, both in terms of over-exploitation of bait species such as the lugworm in England and the bloodworm in America, habitat destruction associated with the digging activities and disturbance of other wildlife, of which overwintering waders are possibly the most vulnerable. It should also perhaps be pointed out here that in freshwater, there has been a move to ban the use of lead fishing weights, which have been implicated in lead poisoning of bottom feeding birds (O'Halloran et al., 1987).

3.2 CONTAMINATION AND POLLUTION

It is valid also to consider the assimilation and transformation of waste organic matter as part of the biological resources of the estuary as this, as opposed to mere physical dispersion and dilution, depends on the proper functioning of the system. It is to the usage of estuaries as waste receivers, from which stems much of the pollution, that we shall now turn.

Broadly speaking, there are five main categories of contaminant commonly discharged into nearshore and estuarine environments, and these groups and their characteristics are summarised in Figure 3.1. There are several properties of wastes which make them of environmental concern, and broadly speaking, the effect they have on the environment depends on the amounts discharged, their toxicity, their persistence or resistance to breakdown or assimilation, and any tendency to bioaccumulation. The scheme shown in Figure 3.1 is ranked according to amounts discharged, but the other properties also are indicated. However, it must be remembered that there is considerable variability within and between groups, and the trends indicated are merely rough generalisations.

Figure 3.1   Classes of contamination and their properties.

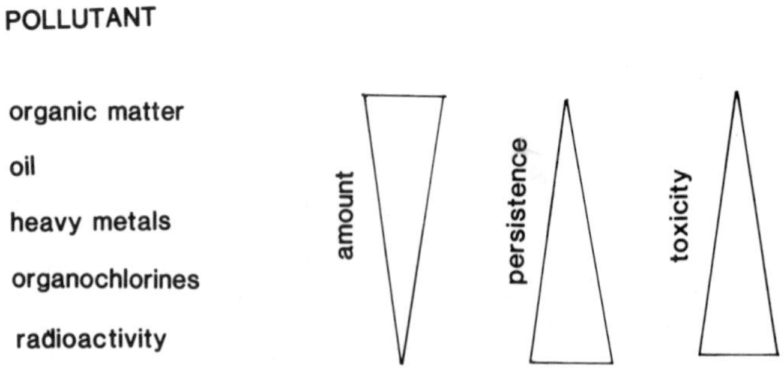

POLLUTANT

organic matter

oil

heavy metals

organochlorines

radioactivity

amount   persistence   toxicity

In terms of the inputs into the North Sea, by far the major contribution is in the form of organic matter such as sewage or sewage sludge. Along with

these are the direct industrial inputs, some of
which again may contribute to oxygen demand, but
whose main contribution would be via substances such
as oils, metals and organohalogens.

An estimate of the inputs of the different
classes of contaminant is shown in Table 3.1 along
with the contribution made directly by anthropogenic
sources.

An analysis of some of these inputs has been
attempted by van Pagee et al. (1986) demonstrating
both the origins of the material and the pathway by
which they entered the system (Figure 3.2). Coastal
inputs, though a small fraction of the total water
mass, account for between 30% and 50% of the total
input, but it is worth noting that atmospheric
deposition was responsible for large amounts of
notably lead and zinc.

Table 3.1 Inputs into the North Sea (data from
various sources)

Contaminant	Load (1000 t/y)	% Anthropogenic
Sewage	1800000	100
N	3049	27
P	369	30
Zn	31.3	44
Pb	6.3	34
Cu	7.8	31
Cr	8.1	24
Cd	0.5	35
Hg	0.09	48
Oil	194	?
Organochlorines	0.2-0.3	100
Radioactivity	?	2

By country of origin, the Dutch input seemed
particularly large until it is pointed out that this
was due to the enormous quantities of water in the
rivers which discharge through the Dutch coast
(Figure 3.2). Typical of these is the Rhine, which
is often not so humourously referred to as the
"sewer of Europe", and collects the wastes from each
of the countries it passes through before
discharging it at the Dutch coast.

Other workers have suggested categorisation by
function based on the other properties such as
persistence or toxicity. Morel and Schiff (1983)
proposed a classification according to reactivity or
persistence with three grades from easily degradable

to very resistant, but persistence and toxicity do not necessarily go together (despite the indications in Figure 3.1).

Figure 3.2   Contaminant inputs into the North Sea by source.

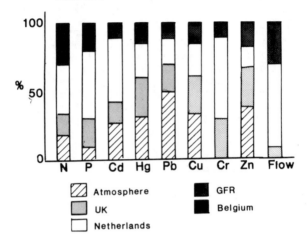

Page (1983) classified contaminants into four categories:

Class 1 - "natural" materials such as organic matter and nutrients which already cycle through the system in large quantities. This category would include all BOD compounds and thermal pollution, and is composed largely of those waste constituents which are easily assimilated and non-toxic.

Class 2 - pathogens such as bacteria and viruses. While these may be rapidly assimilated or at least quickly neutralised by the system, they present a considerable danger to human health. It is however worth noting that the danger is limited to human health, and the rest of the system may survive or indeed flourish in these situations.

Table 3.2 Sources and types of pollutants, ** - primary contaminants, * - secondary contaminants: BOD, suspended solids (SS), nutrients (Nutr), pathogens (Path), heat, hydrocarbons (HC), heavy metals (HM), organics (Org) and organochlorines (OrgCl)

Pollutant

Industry/Source	BOD	SS	Nutr	Path	Heat	HC	HM	Org	OrgCl
Agriculture	**	**	**	*					
Dairy/Meat Processing	**	**	**	*					
Sugar beet/cane	**	**	**				*		
Beverage	**	**	*			*			
Textiles/wool	**		*			*	**		*
Paper/pulp	**	**				*	**		
Plastics						*		*	**
Chemicals						*		**	**
Automobile	*	**				**	**		
Metal (Heavy industry)	*	**			*	**	**	*	
Cement/glass/asbestos		**				*	**		
Electricity generation		*			**	*	**	*	
Petrochemical	*	**		*		**	**	*	**
Harbour/Port activities	**	**	*			**	**	**	**
Petrochemical	*	*							

Class 3 - heavy metals.

Class 4 - toxic chemicals causing carcinogenic, mutagenic or teratogenic effects through their action on the genetic material. Such genotoxic agents include organic solvents and indeed many of the new organochemical compounds, and radioactivity.

Most wastes contain mixtures of the above according to the population and the types and degree of industrialisation in their catchment area. Table 3.2 shows a list of the major and minor sources and amounts of the common contaminant classes discharged by industry into estuaries. The ratios of the different contaminants varies from industry to industry: for example, those involved on the dairy/food side pose problems primarily of BOD, whereas the traditional heavy industries also have substantial inputs of the persistant pollutants.

The tendency is now for new industries, particularly those with "difficult" wastes to install separate treatment works, as recovery of contaminants, whether for economic or environmental reasons, is much easier from a small volume of concentrated effluent. Removal of toxic substances at this stage also prevents poisoning of biological treatment works and keeps open many more options for the disposal of the settled sludge.

Table 3.3   Constituents of a "typical" effluent (primarily domestic waste with a small industrial input). (Data from several sources: mg/l unless stated)

BOD	200
DOC	100
POC	200
Suspended solids	250
Total N	10
Ammonia N	5
Total P	1.2
Oil (PHC)	0.5
Zinc	0.4
Lead	0.05
Copper	0.05
Cadmium	0.01
Mercury	0.001
Total coliforms	10^6/ml
Faecal coliforms	10^5/ml
Faecal Streptococci	10^4/ml

A typical effluent will contain a wide range of substances (Table 3.3). Ater this has been discharged into the estuary there is an initial dilution, the degree depending on the size of the respective water bodies and on the amount of mixing in the system, which then begins to deal with this new cocktail.

The fate of the mixture is depicted diagrammatically in Figure 3.3. Those elements in the lower half, namely oils, metals etc., will be dealt with separately under their individual headings, and we shall now look at what happens to the organic matter, the nutrients and the pathogens in some more detail.

Figure 3.3 Fate of wastes.

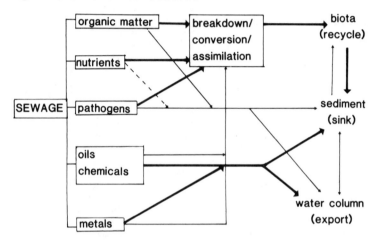

3.3 SEWAGE AND ORGANIC WASTES

Organic loadings are commonly expressed as a function of the oxygen consumed by their assimilation into the system, hence the practice of referring to them in terms of Biochemical (or Biological and Chemical) Oxygen Demand (BOD). The accepted practice is to measure the amount of oxygen consumed in the dark (to prevent generation of oxygen by photosynthesis) at a constant temperature for a stated period, which for historical reasons are 20°C and 5 days respectively.

There are quicker methods to estimate oxygen demand such as using shorter incubation times, but

the shorter the period, the greater the effect of
the lag phase, due to the kinetics of bacterial
growth in the bottle, and the greater the correction
factor required to convert to 5-day BOD. Ademoroti
(1984) has calculated factors of 1.10, 1.243, and
1.58 for incubation periods of 4, 3, and 2 days
respectively, with close agreement between the
correction factors for a typical community waste and
an industrial (brewery) waste.

While there are drawbacks to this method, all
others, such as TOC (Total Organic Carbon = DOC +
POC) measurements, which with the proper equipment
are much more rapid and so are used in situations
such as on the spot monitoring of effluent quality
as it is discharged, are usually referred back to
the standard 5-day BOD and calibrated against it.

The actual BOD of any effluent will vary
according to the types and quantities of the inputs,
and likewise the relationship between BOD, COD and
TOC will vary. Ademoroti (1986) showed that the
ratio between BOD and COD varied from around 0.4 to
0.8 and that TOC varied from 1.00 to 1.85 of BOD.

As was seen in Chapter 2, there already exist
major pathways for the assimilation of POC and DOC
into the estuarine system. Such extra inputs have
then three possible fates (Figure 3.3), namely
conversion into additional living biomass,
incorporation and storage in the sediment, or export
to the sea. Obviously the preferred outcome is
incorporation into the system, as the other two are
at best short-term solutions. If the organic matter
builds up in the sediment, then a point will be
reached when sediment oxygen consumption, due to
microbial breakdown of the organic matter, exceeds
supply, the sediment becomes anoxic and the
biological and chemical character of the system
changes abruptly. Export to the sea can work
provided again that the receiving system can cope,
but it is really just transferring the problem to
another, albeit somewhat larger system. It is rare
for the export option to be effective enough to
prevent damage to the estuarine system, and in most
cases the sea simply acts as an overflow when the
estuary is overloaded or the system has crashed.

3.3.1 Nutrients

There are however problems associated with the
breakdown and assimilation of organic matter into
the system, and one of them as has been mentioned is

that the process consumes oxygen: excess consumption
then leads to problems of anoxia. The second problem
is that in the process of breakdown nutrients such
as N and P are released. Again, as with carbon,
there are established pathways that the system can
use to deal with this extra input, and the options
are much the same (Figure 3.3). The problem may
arise in those situations where one of the
nutrients, generally N, has been limiting primary
production, and when the limiting factor is removed,
the nutrients are utilised in the photosynthetic
fixing of carbon dioxide and the production of
organic matter as plant biomass.
        This extra plant biomass can affect the
functioning of the system in two ways: firstly, when
there is plenty of light during the day there is
plenty of oxygen produced, and the water column may
actually become supersaturated with oxygen (DO
readings in excess of 100%), but during the night,
oxygen is consumed in respiration at a greater than
usual rate, and may accelerate the tendency of the
system to anoxia. Secondly, relatively little of the
primary production is consumed directly (see Chapter
2) and so the excess production enters the food
chain via the decomposer cycle, giving a situation
not unlike that caused by direct organic matter
loading. Riley and Chester (1971) calculated that
for a sewage effluent with a total N content of 40
mg/l, the resultant phytoplankton biomass would have
a potential BOD of just under 800 mg/l, in other
words almost four times an average direct BOD load
(see Table 3.3). Oxygen depletion from excess
nutrient loadings have led to massive fish kills off
New Jersey and there has been a $60 million loss to
the commercial clam fishery alone (EPA, 1987).
        The excess algae can be a direct nuisance, as
algal mats on beaches are unsightly and deter
bathers, and to many people are an outward and
obvious sign of pollution. The link between algal
blooms and nutrient input has been known for the
best part of a century (Parry and Adeney, 1901), yet
there are still difficulties in establishing a
direct link between sewage input and algal mats
(Soulsby et al., 1985b). These difficulties arise
partly from the routes by which the nutrients enter
the system, since a significant part of nutrient
input is from diffuse rather than point sources,
especially where the river drains an agricultural
hinterland, and partly from the fact that the
estuarine sediment acts as a sink for nutrients,
such that algal growth and nutrient input need not

be directly correlated at any one point in time.

Phytoplankton blooms (sometimes called red tides although in actual fact the colour varies) discolour the water with the same effect, and in addition can often result in skin rashes or allergic complaints from those who come into contact with them. Some blooms, as will be mentioned later, produce toxins which can be accumulated in shellfish with severe or even fatal effects on the consumer.

The tendency of both organic matter and nutrients to accumulate in the sediment sink often obscures any direct relationship between input and effect, for example by introducing a lag phase in which stored material does not become available or is not utilised until there are favourable environmental conditions such as higher temperatures to promote microbial activity or sufficient light for photosynthesis. However, this tendency of the sediment to act as a sink does mean that the system possesses a mechanism by which variations in supply and demand can be to some extent evened out and as such it performs a valuable function in the preservation of some sort of equilibrium in the system.

3.3.2 Pathogens

The amounts and types of pathogens in discharges vary with the health of the population in the catchment area. While some such as the coliforms and others associated directly with faeces may occur in relatively constant concentrations, those associated directly with medical conditions ranging from nematode worm infection to viral diseases can only be introduced from existing human reservoirs (although in some cases there may also be an intermediate host).

If man were to be excluded from the estuarine system, then the fate of pathogens would not have to be dealt with any differently from organic matter, ending up incorporated into the biota or the sediment or exported (Figure 3.3). However, until they are killed or de-activated, they remain a potential hazard to man not only through contact with contaminated water or sediment, but also through swallowing water or the ingestion of organisms which have themselves used the pathogens as a food source. Leptospirosis is a contact infection caused by the entry of a bacterium through abrasions in the skin or through the mucous

82

membranes. In the United States, Leptospirosis
outbreaks, of which about 100/year are reported
(Laws, 1981), are confined to the summer months and
are associated with bathing in contaminated waters.
It is further estimated that there are a similar
number of viral gastroenterology cases a year in New
York from the consumption of contaminated shellfish.

The principal culprits in the latter case are
the bivalves, which obtain their nourishment by
filtering out food and organic particles (making no
distinction between pathogens harmful to man and
their regular diet). As a bivalve such as an oyster
may filter several litres of water a day, the
capacity to concentrate microbes (and incidentally
chemicals or other substances) is considerable. Only
a small fraction of the microbes filtered out of the
water are actually digested, and the rest pass
unchanged through the gut (Birbeck and McHenery,
1982) and remain potentially harmful both while in
the gut of the bivalve and after defaecation.

The bivalves can be purified by allowing them to
filter for a few days in purified water, by which
time all the pathogens in the gut have been passed
out in their faeces. They can also be made safe to
eat by cooking for a sufficient time at high
temperature to kill the pathogens, but occasionally
another complication arises. Paralytic shellfish
poisoning, or PSP for short, occurs when the bivalve
has not only concentrated the phytoplankton
involved, but also the toxin they produce. When this
plankton is present in large numbers, then lethal
doses of the toxin can be accumulated, and the toxin
is not denatured or deactivated by cooking or
processing.

The PSP toxin seems to act by blocking the
supply of sodium to the nerves, whereas the NSP
(neurotoxic shellfish poisoning) toxin has the
opposite effect, causing the nerves to fire
continually. A less extreme form is DSP (diarrhetic
shellfish poisoning), which though unpleasant is not
usually fatal, but in all cases, the causative
organism need not necessarily be present in the
water in great numbers (tens or hundreds per litre)
to produce the toxic effect.

By and large, the pathogens are adapted for life
in the human gut, and so estuarine conditions are
decidedly non-favourable. Nevertheless, the rate at
which they die off varies according to a number of
factors, including sunlight, temperature, salinity
and starvation as well as on the physiological state
of the organism itself (Gameson, 1985).

The mortality rate in sunlight is approximately two orders of magnitude greater than that in the dark, with a lesser effect due to temperature, and die-off at 25°C is around one order of magnitude greater that at 0°C. In estuaries, the dilution of the sea water by fresh water may decrease mortality rate, but a much more important factor is likely to be the turbidity, which by severely restricting light penetration will drastically reduce mortality rate.

The actual rates at which the different pathogens die off depend on the organism involved, with the most resistant forms being those capable of producing spores, such as *Clostridium perfringens*. Commonly, coliform mortality is more rapid than many of the other forms such as *Salmonella* or the Streptococci, although experimental data is not always in completete agreement (Gameson, 1985) and considerable anxiety surrounds the problem of pathogen survival (especially viruses) and the problems of deciding on the criteria for modelling or management (Gray, 1986).

Treatment of the effluent before discharging will reduce the amounts entering the system. Primary treatment, involving maceration and settlemnt only will typically reduce the BOD input by something like 30%, while the addition of secondary treatment and biological digestion or filtration will bring it down to about one-third of the original BOD load depending on the retention time of the treatment. In situations where nutrient loading is to be avoided at all costs, there may also be added a third stage (tertiary treatment) of chemical precipitation, but this usually applies only where the receiving water body is very restricted, such as a small river or lake. The concentration of pathogens, as with BOD, decreases roughly according to the degree of treatment, and according to the retention time of the treatment system. Sedimentation reduces the concentration by between 40% and 70%, while activated sludge treatment or biological filtration will reduce the numbers by between 60% and 99%, including viruses (Pike, 1975). Sterilisation by peroxide, chlorination or UV is the only way to ensure complete safety, albeit at considerable cost.

Chlorination is more commonly used as a biocide for industrial cooling waters, to prevent blockage of pipes by fouling organisms. While screens on the intake will keep out the larger organisms such as fish, the conditions inside many of these systems are ideal for the settlement and growth of marine

invertebrates. Chlorine can of course be produced by
electrolytic action from seawater, and dosages can
be either continuous or intermittent. In either
case, the concentration at discharge should not
exceed an average of 0.2 mg/l and 90% of this decays
back to chloride within 30 minutes. One note of
caution to be sounded here is that some halogenated
organics (which will be dealt with later with the
organochlorines) are produced, and the fate of these
and their long-term effects are not yet well
understood (GESAMP, 1984a). Chlorination of an
effluent, while rendering it sterile, also tends to
sterilise the area round the outfall, and
chlorinated effluents have a much more severe impact
on the system and the biota (see also Chapter 5).

3.3.3 Sewage sludge

Most estuarine discharges receive at best
secondary treatment, and the common practice is to
treat only as far as the primary stage, if only
because the removal of the larger constituents makes
it easier to handle hydraulically. Treatment to any
stage, while it does remove some load from the
effluent results in the production of a sludge to be
got rid of somehow. If this is disposed of in the
same water body as receives the effluent, then the
problem has only been transferred, with two point
sources instead of one. Sludge dumping sites are
customarily at the mouth of, or just outside
estuaries, but the disposal of the sludge, both in
terms of its constituents and quantities, is
irrevocably bound up with the whole question of the
type and degree of treatment. Indeed, as will be
discussed in Chapter Five, the costs of the sludge
disposal may be used as an argument for a lesser
degree of treatment.
From the point of view of the system, there is
little distinction between the two, effluent or
sludge, except the form in which it arrives. Because
settled sludge has a much higher solids content, and
because it tends to arrive in barge-loads at the
bottom, there is often a physical blanketing effect,
but otherwise the fate of the constituents is the
same as shown in Figure 3.3.
Physical blanketing may also result from the
settlement of suspended material, either through
direct sedimentation or flocculation, and while some
organisms may be able to cope with the input, others
may be smothered or deterred by it. This effect has

particularly to be noted where there are fishing or
spawning grounds in the estuary.

Sludge dumping sites may be characterised as
either dispersive or containing. In the former, the
site characteristics are such (high tidal or other
currents) that the sludge is fairly rapidly
dispersed over a wide area. In the latter, the
dumped sludge tends to accumulate in the one place,
and dispersion away from the site is minimal. A
dispersive site is to be preferred for the majority
of sludges, as it offers the best chance for the
material to be assimilated into the system, and
minimises local damage at the dumping site. A
containing site may only be indicated where the
sludge contains substances that need to be kept out
of the system - heavy metals for example - and the
dumped sludge is then capped with a layer of clean
sediment. However, should the unwanted substances
start finding their way into the system, for example
from a combination of storm and biogenic action,
there is virtually no way in which it can be
recovered, and re-capping may be only another
temporary solution to be renewed every few years.

The same is true of the sites chosen for the
point of discharge of effluent, and the criteria for
the selection of such sites have been set out in
some detail by GESAMP (1982), and will be discussed
in a later chapter.

3.4 HEAT

It is appropriate here to consider briefly
thermal discharges into estuaries, as, from the
system's point of view, this is quickly and easily
assimilated.

The impacts of cooling water discharges can,
like chlorination, be divided into near-field and
far-field effects. Near-field effects are limited to
the areas immediately adjacent to the discharge,
where for example the rise in temperature imposes a
direct stress on the organisms. Far-field effects
would be experienced in a bay for example, where the
average temperature is raised by only a small
amount, and the biological effects will be minimal.
Some disruption of temperature dependant processes
may occur, for example acceleration of gametogenesis
and earlier maturation, but in most estuarine
systems, there is little evidence of any such
damage. On the contrary, the discharges have often
been harnessed for mariculture, where the higher

temperatures permit more rapid growth and in some cases allow the spawning of non-indigenous species.

Problems with thermal discharges arise only where the organisms in the system may be living at or near the limits of their thermal tolerance. As with other pollutants, the effect may be enhanced if there is insufficient capacity in the receiving waters to absorb this extra input, and so additional caution is needed in regard to thermal discharges in tropical or semi-tropical environments and in enclosed bays or lagoons with little water movement or exchange.

3.5 HYDROCARBONS

Most urban discharges contain some quantities of oil or petroleum hydrocarbons (PHCs) (see Table 3.3), from sources such as road run-off or accidental (and not so accidental) spillages into the sewerage system. Spectacular examples of oil pollution occur after tanker accidents, but in global terms these account for only a fraction (less than 5% (Sasamura, 1981)) of the PHC input into the marine environment. In contrast urban and coastal wastes and runoff account for over 40% (Sasamura, 1981), and these of course go directly into the estuary. In addition, the use of the estuary for shipping and industry means hydrocarbon inputs (see Table 3.2), and the presence of oil installations, whether as extraction, storage or petrochemical works results in contamination in proportion to the amount of oil handled.

The fate of oil spilled or discharged into the marine environment is summarised in Figure 3.4. From the initial point source, the slick spreads out along the surface in the direction of the prevailing wind(s). As it goes, the lighter fractions evaporate off and the heavier fractions start to sink to the bottom. The remainder forms an oil/water emulsion known colloquially as "chocolate mousse" on account of its colour and texture. There is some cycling of hydrocarbons within both the atmosphere and the water column, through aerosol/droplet exchange with the air and uptake and release by organisms in the upper water layers.

There are bacteria in the natural environment capable of breaking down hydrocarbons, but normally these are present in very low numbers. Their action is aided by high temperatures, as with all microbiological activity, and by the substrate surface area available. The oil is also broken down by physical and chemical processes such as photochemical oxidation, with the greatest rates at the sea surface or once the oil is stranded on the shore. Oil will therefore persist much longer in areas of low water energy such as mudflats, particularly in more northerly latitudes.

Figure 3.4   Fate of oil in the marine environment.

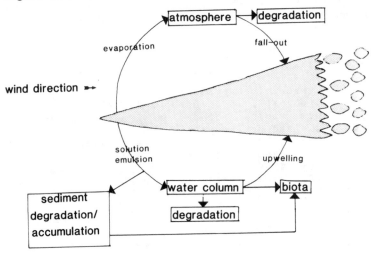

The toxicity of hydrocarbons depends largely on the length of their carbon chains, which determines their fat solubility and the ease with which they are able to pass through the cell membrane, so that the lighter fractions are the more toxic. Roughly speaking, fuel oils and gasolines are toxic around 60 - 200 mg/l, diesel oils at 300 - 3,000 mg/l and lubricating oils at 3,000 mg/l to as high as 20% (Nelson-Smith, 1977). As it is the lighter fractions which are more rapidly lost from spills, it is often the case that by the time a spill comes ashore, the main effect on shore organisms is a smothering effect by the residual heavy fractions. In these situations, the best policy is often to leave the shore to the natural cleaning process, especially on exposed shores, as cleaning with detergents, or even mechanical cleaning can cause serious damage. After the Torrey Canyon wreck off south-west England, it was found that the detergents used and in particular the oil/detergent mixes were many times more toxic than the oil itself (Nelson-Smith, 1977), and less toxic substitutes were developed. In his review, Johnston (1984) pointed out that the tendency nowadays is to recommend as little action as possible, as in the majority of cases, the system recovers more quickly if left to its own devices.

With diffuse inputs, the effect is more difficult to gauge, and since the hydrocarbon must be in the water to be available to organisms, many toxicity studies now quote their findings in terms of the water-soluble fraction (WSF) of the oil in question. Depending on the oil, the WSF may range from 4% to 40% of the total. Like all pollutants, oil will affect the system at levels below those which directly cause mortality (sub-lethal effects), but in addition oil taken up by organisms will cause tainting of flesh and tissues, rendering these unpalatable. For this reason, oil and commercial fisheries are uses best kept apart if possible.

Among the variety of substances present in crude oils are polynuclear aromatic hydrocarbons (PAH's), some of which are potential or actual carcinogens. However, their concentration is so low that even in oil-contaminated seafood, there is little likelihood of damage to human health, and no evidence exists to show that they might be bioaccumulated up the marine food chain (Mertens and Gould, 1979).

Recently, PHC budgets have been estimated for two coastal situations in the Hudson Raritan estuary, New York (Connell, 1982) and in a Danish fjord (Jensen, 1983). In the Raritan estuary, some 80% of the input was attributed to runoff and sewage discharge, but these played a minor role in Kalundborg Fjord which was dominated by the input from refinery effluent (Figure 3.5).

The figure clearly shows that the system is being overloaded and cannot adequately deal with the amounts entering. Despite the relatively large amounts removed by advective water transport, significant amounts are accumulating in the sediment and organisms such as *Mytilus edulis* are seriously contaminated. The amounts with which the system can cope, in other words the amounts degraded in the sediment and the water column, come to just over 10% of the total input, and just over half the amount accumulating annually in the sediment.

Delaune et al. (1980) have shown that little if any degradation of oil takes place in reduced sediments, so that any tendency of an estuarine system to anoxia will seriously affect the rate at which hydrocarbons can be broken down and

89

assimilated into the system.

Figure 3.5   Hydrocarbon budget for the Kalundborg fjord, Denmark. (After Jensen, 1983).

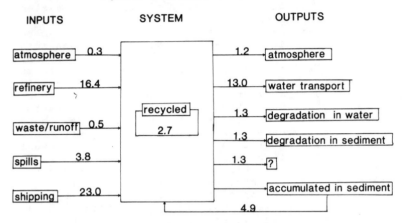

The overall form of Figure 3.5 is basically the same as the nutrient and other budgets in Chapter 2, with the main difference being that the rates of assimilation and the ability of the system to handle the new substances are much less, hence the greater persistence of oil. In many respects, the behaviour and effect of oil in the system is intermediate between the non-persistent contaminants such as organic matter or nutrients, and those in the bottom half of Figure 3.1, the heavy metals, organo-chlorines and radioactive contaminants that we shall now go on to consider.

3.6 HEAVY METALS

Heavy metals are present naturally in the estuarine environment, and in many cases their presence is essential to life itself. Iron and copper are incorporated into blood pigments and others like zinc are co-factors for enzyme activity. The oceans themselves are vast reservoirs of metals, which are continually being lost through incorporation into the sediment and replaced through the weathering of rock. Some idea of the relative scales of natural and anthropogenic input and the amount already present can be gained from Table 3.4.

Table   3.4   Metal   sources   and   amounts.   (From
               Phillips, 1980)

Metal	Sources		Total in oceans $(10^6 t)$
	Natural $(10^3 t/y)$	Man $(10^3 t/y)$	
Iron	25000	395000	4110
Copper	375	6000	4110
Zinc	370	5320	6850
Nickel	300	481	2740
Lead	180	3200	41
Tin	1.5	277	14
Cadmium	-	17	68
Mercury	3	10.5	68

In some cases, nickel for example, anthropogenic
input on a global scale is minimal compared to the
existing reservoir in the oceans. In others, like
mercury, anthropogenic input is a significant
fraction of the total and is adding appreciably to
total ocean content (see alo Table 3.1).

However, metal input, like that of other
contaminants is not distributed evenly over the
oceans, but is concentrated in estuaries. In places,
even the natural input can cause local concern, as
with mercury at certain locations along the Italian
coast where the geology of the river catchment areas
contains mercury-rich ores. Likewise, the use of
estuaries as receiving waters means that many are
subject to considerable metal loadings.

The toxicity of heavy metals varies according to
the metal involved, its form, and the target
organism, and this is illustrated in Table 3.5.

Table 3.5   Toxicity of selected heavy metals (µg/l)

Metal	Toxicity Range	Reference
Arsenic	3,000 - 60,000	Taylor (1981a)
Lead (inorganic)	1,000 - 100,000	Taylor (1981c)
Lead (organic)	0.02 - 300	GESAMP (1984b)
Zinc	200 - 20,000	Taylor (1981d)
Tin (inorganic)	100 - 1,000*	GESAMP (1984b)
Tin (organic)	0.005 - 0.05	GESAMP (1984b)
Copper	20 - 100,000	Taylor (1981b)
Cadmium	0.1 -50	GESAMP (1984b)
Mercury (inorg)	5 - 4,000	Taylor (1977)
Mercury (org)	0.2 - 8,000	Taylor (1977)

*limit of solubility 35 µg/l (GESAMP, 1984b)

These values relate principally to metals in the ionic form, in which they are most available for uptake, and the availability of other forms will be discussed shortly. Note however that the organometals are on average more toxic than the ionic forms, and this is even more pronounced in metals like tin, which in itself is only moderately toxic, but when incorporated into tri-butyl tin (TBT) results in an extemely effective and powerful biocide used in antifouling paints. In this form it was shown, not unsurprisingly perhaps, that it not only kills or discourages organisms from colonising boat hulls, it also has severe effects on the whole system. Molluscs in particular seem to be extremely sensitive and effects on some species can be detected at levels as low as 0.02 µg/l (Gibbs and Bryan, 1986).

Metals have also a tendency to bioaccumulation. This term is often used to describe two different processes, more properly called bioconcentration and biomagnification respectively (see Table 3.6). The former is the ability of organisms to accumulate contaminant loads greatly in excess of those in the ambient water. This property is shared by the sediment, and because it makes detection of trace contaminants much easier, is often used to justify their choice in monitoring programmes. However, the relationship between water metal levels and sediment or organism levels is by no means straightforward, even when the metal is presented directly in ionic form (Wilson, 1983).

Table 3.6 Mercury and methyl mercury: bioconcentration and biomagnification via the food chain. Levels (µg/g) and concentration factor (CF)

Parameter	Inorganic Hg µg/g	CF	Organic Hg µg/g	CF
Water	0.00002	1	0.000005	1
1y producer	0.05	2×10^3	0.001	2×10^2
1y consumer	1.77	8×10^4	0.20	4×10^4
2y consumer (muscle)	0.32	2×10^4	1.04	2×10^5
(liver)	0.96	4×10^4	1.57	3×10^5
3y consumer (muscle)	1.75	8×10^4	2.60	5×10^5
(liver)	5.55	3×10^5	6.25	1×10^6

Biomagnification, on the other hand is the concentration of substances up the food chain, such that relatively low and innocuous levels at the bottom of the chain are progressively accumulated to harmful or lethal dose levels in organisms at the top.

Biomagnifaction via the food chain is of direct relevance to man, as man and other vertebrates such as birds or seals are at the apex of many estuarine food webs. The best known case involved mercury in Japan at Minimata, where the metal was discharged into the bay, concentrated up through the food chain to the fish which constituted a large part of the diet of local fishermen and their families. Altogether 107 fatalities were recorded, with a further 800 verified and 2800 unverified victims. A similar case was recorded, again in Japan, with cadmium in the so-called "Itai-Itai" disease. Here however, the evidence was by no means so clear cut, and diet deficiency was also a significant factor, along with combinations of other metals particularly aluminium and zinc.

Mance (1987) has recently reviewed the evidence for both phenomena and has concluded that while bioconcentration has been consistently shown over the whole spectrum of metals, evidence for bio-magnification is distinctly lacking. Some studies have seemed to indicate that arsenic and mercury could be biomagnified up the food chain, although even here the evidence is inconsistent, and it has been suggested that this process is due to these two elements' affinity for organic matter (Mance, 1987).

The uptake of metals by estuarine organisms varies considerably depending on environmental and other factors. These factors and their effects have been reviewed at some length by Phillips (1980) particularly in relation to the suitability of organisms as indicators of pollution, and the criteria which indicator organisms should satisfy are discussed in the following chapter. Bryan (Bryan et al., 1980; 1985) has investigated in detail the relationship between selected metals and estuarine organisms. The most important factors controlling metal levels in sediment infauna are the actual concentration in the sediments and the partitioning of these metals, but these two factors may only explain around 50% of the total variability, with a very diffuse relationship in cases such as Zn where there is regulation by the organism (Luoma and Bryan, 1982).

The events at Minimata emphasised the dangers of metal comtamination in that relatively low discharge or environmental levels could be greatly magnified by the system, and a great deal of attention has been subsequently paid to the mechanisms by which this accumulation, in both senses, is achieved.

The fate of metals discharged into the estuarine environment depends on their affinity for their various ligands. Some 85% of cadmium introduced into an experimental mesocosm remained in solution, and the major removal pathway was via the water column, while the opposite was true for lead, of which 90% was transferred to the particulate phase (Loring and Prosi, 1986).

The movement of these particles is in turn governed by the hydrodynamic regime, and they may then remain in suspension, and the metals can cycle through the water column, or they may settle out and be incorporated into the sediment. Just as organisms can accumulate metal levels greatly in excess of those in the water column, so can sediment achieve concentrations several orders of magnitude greater than water. Metals are removed from solution by a mixture of absorption, either by free organic matter or the organic coating on sediment particles, and adsorption of the metals directly onto the particle surface. Since sediment organic matter and hence absorption capacity, is strongly correlated with particle size, and surface area for adsorption likewise depends on particle size, metal concentration usually but not invariably varies with the particle size distribution of the sediment (Wilson et al., 1986).

The availablity of metal from sediments to organisms depends on the partitioning, as mentioned above, and this can be ascertained by differential extraction, in which progressively stronger agents are used to leach out or release the metals from their ligands. Different metals have different availabilities according to their binding states (Figure 3.6).

An important point to note from Figure 3.6 is that the metals are much more available from oxidised sediments, with rapid transformations of zinc and cadmium especially from the sulphidic/ organic binding to weakly-held and easily reducible species.

These interactions of sediment, metal and oxygen have drawn a great deal of attention to the turbidity maximum in the estuary. This area is the focus of the sedimentation/ deposition regime and in addition has strong salinity and oxygen gradients. Within this zone and within the course of its

passage up and down the estuary with the tide, there
will be considerable mobilisation and flux of metals
between the sediment and the water column, and the
movement and availability of manganese for example
is controlled largely by sediment fluxes (Morris et
al., 1982).

Figure   3.6   Metal   partitioning   in   estuarine
                     sediments: a) oxidised; b) reduced.
                     (After   Kerstner   and   Foster, 1986;
                     Martin et al., 1986).

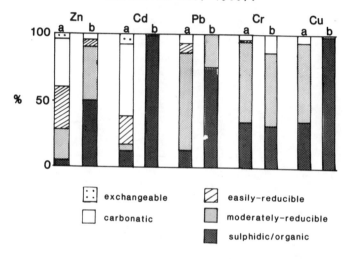

The role of the turbidity maximum and the
relationship with total contaminant concentration is
demonstrated in Figure 3.7, which shows the
behaviour of cadmium in the River Weser estuary in
Germany (see also Figure 2.7).

As with the other contaminants, the sediment
plays a major role in the estuarine budget, although
only a small fraction of the total metal content of
the sediment may be available. Elliott and Griffiths
(1986) have quantified the standing mass of mercury
in the Forth estuary in Scotland, and have shown
that the sediments alone account for 97% of the

total. The bulk of the remainder (2.7%) was present in suspended particulate matter with insignificant amounts in the soluble phase and the biota. From the distribution, the suggestion was that the critical pathways for the entry of mercury to the system were from the direct accumulation by infauna from sediments and the uptake of suspended sediments by filter feeders (Elliott and Griffiths, 1986).

Figure 3.7   Partitioning of Cd down the Weser estuary: a) % particulate; b) total Cd in particulate phase (shaded) and dissolved phase (unshaded). (After Calmano et al., 1984).

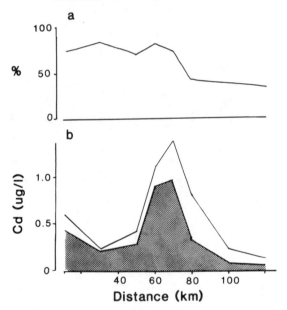

The formation or presence of organometallic compounds may increase both the accumulation (Table 3.6) and the toxicity (Table 3.5), but little is known, apart from a few selected substances, of their fate or behaviour in the system.

Organometals may be both natural and man-made in

origin. Inorganic tin may be gradually biomethylated
to mono-, di-, tri- or tetramethyl tin, and although
this natural organotin is biodegradable, the
synthetic compounds are more durable and require
extremes of acidity or alkalinity to break them down
(GESAMP, 1984b). In the past few years organotins
have increased in importance to rank fourth in the
production of organometals. They are used as
catalysts and stabilisers as well as biocides, and
production is currently estimated at around 50,000
tonnes/year (GESAMP, 1984b). While this is
insignificant in relation to the total tin budget of
the oceans, the inputs are concentrated around
coastal waters (and estuaries in particular), they
are extremely persistent, they can be bioaccumulated
and they are extremely toxic. Thus they fulfil
almost all criteria of pollutant danger, and are a
source of increasing interest and concern.

3.7 ORGANOCHLORINES

Many of the properties of organometals are
shared by the next category of contaminants to
consider, the organochlorines, and also by many of
the synthetic organic compounds now being produced,
directly or as by-products, by industry. These
compounds are sometimes lumped together under the
heading of refractory organics, that is they are
extremely resistant to degradation. Their effect on
the system not only derives from this persistence,
but also from their tendency to accumulate in body
tissues, in which their affinity for fatty tissues
and lipids enhances their mobility and build-up, as
well as from their often carcinogenic or mutagenic
properties. An example of such compounds has already
been mentioned, the PNA's associated with
hydrocarbons, but in general this class, excluding
the pesticides and PCB's, is present in the
estuarine environment in barely detectable traces,
even in those situations where there are known
direct inputs (van de Meent et al., 1986).
When organochlorine pesticides were first
introduced in the 1940's they were hailed as miracle
workers since they were not only very effective
against their targets but also there were no
indications of a wider toxicity to other organisms.
However, this situation did not last, and evidence
of damage particularly at the top end of the food
chain forced a re-think on their widespread and
indiscriminate use. The publication in 1962 of

Rachel Carson's "Silent Spring", in the teeth of
opposition from both the pesticide industry and
Government Departments brought the issue to public
attention and was one of the prime factors in
forcing action to change the situation.

As their name suggests, organochlorine
pesticides consist of one or more chlorine atoms
attached to a hydrocarbon. The number of chlorine
atoms and their location is fundamental to both the
toxicity of the resultant compound and its ability
to resist degradation in the environment. As a
general rule, the more chlorine substitutions, the
more toxic and more refractory the chemical.

The best-documented of these chlorinated
hydrocarbons is the insecticide known, like the
others by its abbreviated name DDT rather than by
its chemical name (1,1,1-Trichloro-2,2-bis(p-
chlorophenyl)ethane!) from its structure (Figure
3.8). In the environment, DDT is broken down to DDD
and DDE (Figure 3.7), and although both the
degradation products resemble the parent in toxicity
and persistence, there is evidence to suggest that
DDE is the worst of the three in this regard.

Figure 3.8   DDT, DDD and DDE   structure.

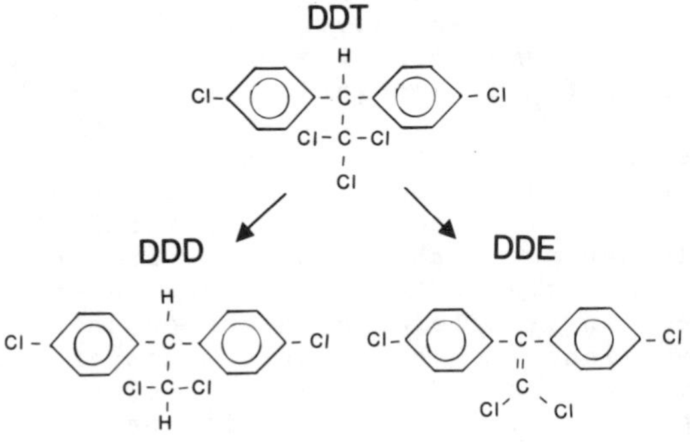

Another group of chemically similar pesticides,
dieldrin, aldrin, chlordane and heptachlor have all
been banned by the United States EPA on the grounds
that they are carcinogenic, although their use is
still widespread in other countries.

Although not pesticides, the chemical similarity
of polychlorinated biphenyls (PCBs) to the
chlorinated hydrocarbon pesticides means that they
may be considered together as far as their effect

and fate in the system are concerned. The principal industrial uses of PCBs are in plasticers for waxes and in various fluids such as transformer fluids, hydraulic fluids, capacitor dielectric fluids and lubricants. Again, while greater control has been exercised in the manufacture and discharge of these substances in the past ten years or so, they are exceptionally stable compounds capable not only of persisting for many years after introduction but also of dissemination over wide areas. Table 3.7 gives a list of pesticide and PCB concentrations found by Wilson and Earley (1985) in seabirds on the coast of Ireland. While the concentrations are less by a factor of one or two than comparable situations in mainland Europe or America, they do illustrate both the capacity of these substances to travel from their source and the ability of organisms to accumulate them in their tissues. It is worth noting also that the levels of DDT were in general lower than those of its metabolite DDE, showing that although the use of DDT and hence its introduction to the environment has decreased dramatically, the effects will linger on for many years in the form of the breakdown products.

Table 3.7 Pesticide and PCB levels (ng/g) in shag and cormorant eggs off south-west Ireland (data from Wilson and Earley, 1985)

Organochlorine	Cormorant		Shag	
	Max.	Mean	Max.	Mean
pp-DDT	2.6	1.7	11.7	9.1
pp-DDD	-	-	6.3	5.3
pp-DDE	76.5	52.3	97.7	82.8
op-DDE	0.9	0.6	-	-
Lindane	2.5	1.4	3.1	2.0
Dieldrin	3.0	2.1	39.1	29.3
Endrin	1.6	0.3	-	-
-BHC	1.8	1.3	2.1	1.7
Heptachlor	1.8	1.2	-	-
Quintogen	2.4	1.4	0.06	0.03
Total pesticides	75.7	50.3	177.2	156.8
PCBs	717.9	492.2	1526.6	1234.9
Total pest + PCBs	793.3	548.5	1709.9	1401.8

The ability of these compounds to persist is a result of their novelty in the environment, and the paucity of bacteria or other agents able to use them as substrates - Lindane is one of the few to be readily metabolised. Estimates for the half-life of

DDT range from about five to fifteen years, Dieldrin about the same and Chlordane two or three times longer (Mitchell, 1974). Some PCBs are more refractory still and have been calculated to be up to 300 times more persistent in the marine environment than DDT (Harvey et al., 1974). In contrast, the organophosphorus group of biocides decay within a matter of days on contact with the environment, and so are increasingly being advocated for application despite their initially greater toxicity.

Because of the affinity of organochlorines for fats and lipids, they concentrate in the body in these tissues. In many cases they act directly on the central nervous system (CNS) and much of the concern about their action has centred on the effect that high concentrations in the brain or CNS have on the behaviour patterns. In addition DDE in particular has been linked to thinning of eggshells in birds, and the cumulative effect is not necessarily mortality of adults, but rather a gradual decrease in population through reproductive failure due to disrupted breeding behaviour or non-viable eggs.

In contrast to the somewhat equivocal situation with metals, organochlorines display a marked tendency to accumulation up the food chain (Table 3.8), reinforcing the concern not only for birds, but also for mammals such as seals, which in enclosed coastal areas such as the Dutch Wadden Sea and the Baltic appear to be going through a reproductive decline similar to that alluded to above (Klinowska, 1986). No food chain effects have been detected in humans, even in groups such as fishermen with relatively high PCB intake, but PCB and chlorinated dibenzofurans were implicated in a case of poisoning with contaminated rice oil which resulted in 29 deaths and 1291 other reported victims (Martin, 1977).

The reasons for the lack of evidence for damage to man are unclear: this may partly be due to the lower intake compared to the other organisms mentioned and partly to the species difference in tolerance and also physiology. Because organochlorines have such affinity for fatty tissue, energy reserves in the form of stored lipid may accumulate large amounts (for this reason levels are normally quoted per unit lipid rather than per unit body weight), and when these reserves are mobilised, during a severe winter or for reproduction, the stored organochlorines are then liberated into the circulation. This not only exposes the animal to a sudden high concentration, and it may be already under the stress which led it to metabolise the lipids in the first place, it also means that any high lipid tissues synthesised, such as the yolk in eggs, will inevitably carry with them high residues and hence the link with reproduction.

Table 3.8   Bioaccumulation of DDT (ng/g lipid) and concentration factor (CF). (Phillips, 1980)

Organism	Concentration	CF
Phytoplankton	40 - 100	1
Zooplankton	? - 154	1.5
Flying fish	180 - 1480	15
Cod	1500 - 2500	25
Seal	20000 - 40000	400

Concern over the levels of PCBs in the sediments, a product of years of waste dumping from electrical industries, has led to the closure of over 28 square miles of lobster grounds in New Bedford, Massachussetts, leading to losses estimated at around 2 million dollars to both the commercial and the sport fishery (EPA, 1987). Local beaches have also been closed, with an estimated revenue loss of close on 15 million dollars, and real estate values in the area have declined.

3.8 RADIOACTIVITY

In contrast to the literally thousands of potentially dangerous organochemicals being introduced to the marine environment, radioactivity seems at first to be relatively simple, with only four mechanisms of direct importance, namely alpha (α) and beta (β) particles, neutrons and gamma (γ) rays.

Alpha particles are released when a radionuclide decays, and consist of two protons and two neutrons. Because of their large size and charge, alpha particles penetrate the shortest distance into matter, travelling only a few centimetres in air and stopped by a piece of paper or skin. However, since it is stopped so quickly, the ionising damage to local tissues may be considerable, although those a short distance away may be unaffected. Thus an alpha particle is of little importance as an external radiation source, but can be of great concern as an internal source, after ingestion for example.

Beta particles are much smaller, with a mass equivalent to one electron, although the beta particle itself may be negatively or positively charged. Due to their smaller size and lesser charge, beta particles may penetrate much deeper into tissue, up to several centimetres, and thus spread their damage over a wider area. Nevertheless, although beta particles may travel several metres through air, they are of external concern only in the immediate vicinity of a beta-emitter, and as with alpha particles, the danger is more through internal exposure from ingestion. The chief difference in effect between alpha and beta particles is therefore that in the latter case the damage is less intense, but spread over a longer distance.

There are a few radionuclides which emit neutrons on decay, but the primary source is the high energy neutrons produced by nuclear reactions. The neutrons have a mass of one quarter that of an alpha particle, but of course carry no charge. The lack of charge means that neutrons are not directly associated with ionising damage, but on their passage through matter, they may induce radioactivity by exciting atoms from their non-radioactive states.

Gamma rays are frequently emitted when radionuclides decay, and this high energy electromagnetic radiation is highly penetrating, often passing through an organism without causing any damage at all. However, the damage that does occur is spread over a long path distance, although the intensity is less than that of alpha or beta particles.

The carcinogenic potential of a radionuclide varies with a number of factors, including the target tissue or organism, the location of the source and the amount and type of emitted radiation.

The activity of a radionuclide is measured by the number of atoms that decay, and hence emit, per unit time and one becquerel (Bq) is equivalent to one disintegration per second.

In 1986 a total activity of around 18,000 Bq was licensed for discharge into UK coastal waters, of which around 4,000 Bq was actually discharged (Hunt, 1987). The bulk of the discharge (over 60%) was in the form of tritium (^{3}H), which is principally a β emitter. Around 40% of the establishments licensed discharge directly into estuaries, with nearly all the remainder going into nearshore or coastal waters, and the fate of the substances discharged has been closely monitored from the standpoint of danger to human health for almost twenty years (Hunt, 1987).

On decay, the radionuclide may emit different rays or particle with different properties, as outlined above. The differences in damage likely to be caused are allowed for by a quality factor (Q) as shown in Table 3.9.

Table 3.9   Q factors for selected radiation (ICRP, 1977)

Type of radiation	Q
gamma rays, beta particles	1
neutrons	10
alpha particles	20

Finally, as well as emitting different particles, different radionuclides may emit the same particle but with different energies. For example ^{32}P emits beta particles with a maximum energy of 1.71 meV compared to ^{14}C beta particles with maximum energies of 0.16 meV. Thus one beta particle of ^{32}P would be expected to result in roughly ten times as much ionisation in a tissue as a particle from ^{14}C, and the SI measure of this absorbed radiation energy is the gray (Gy). One gray is defined as the absorbed dose of one joule per kg of matter, and the product of the dose in grays times the Q factor (Table 3.9) is known as the dose equivalent in sieverts (Sv).

The dose equivalent to the critical consumer group from Sellafield discharges into the Irish Sea has been calculated as 0.34 mSv/person/year, which falls within the ICRP limit of 1mSv/person/year and is around one-sixth of the dose from natural sources of radiation in the UK of around 2mSv/year (Hunt, 1987).

In addition to the radiation danger itself, some radionuclides are of special concern because of the way in wnich they can be concentrated or accumulated by sediments or organisms or by the way in which they behave in the organism due to their chemical properties. Table 3.10 lists some of those of greatest concern along with the causes for that concern.

Because some elements are in great demand by organisms, e.g. nutrients for plant growth, calcium for bone and so on, mechanisms have been developed to obtain them preferentially. When a radioisotope comes along that fulfils the required criteria, either by simply being a radioactive version of the element or by being sufficiently similar in atomic structure, then that is sequestered from the environment. The problem is further exacerbated by the fact that in many cases the element is transported within the body to selected sites, thus effectively ensuring its concentration manyfold not only from the environment to the body but also within a small part of the body itself. The mechanism for uranium and plutonium is somewhat different, as these have great affinity for sediments and particulates, i.e. they behave like any heavy metal and the entry into the body is through mechanical as opposed to chemical pathways.

Table 3.10 Bioaccumulating radionuclides

Radio-nuclide	Half-life	Mode of action
^{3}H	12.3 y	As ordinary H e.g. water
^{14}C	5.8×10^{3} y	As ordinary C e.g. food
^{32}P	14.3 d	As ordinary P, in bone and in nutrient cycle
^{90}Sr	28y }	Mimics Ca, accumulated
^{226}Ra	1.6×10^{3} y}	in bone
^{137}Cs	30 y }	Mimics K in body
^{129}I	1.7×10^{7} y}	As ordinary I, thyroid
^{131}I	8.0 d }	accumulation
^{238}U	4.5×10^{9} y}	Particulate uptake,
^{239}Pu	2.4×10^{4} y}	concentrated in lungs

Some indication of the ability of organisms, in this case *Fucus vesiculosus* , to accumulate Pu is shown in Figure 3.9.

The residual currents in the Irish Sea are northward, and a similar northward drift has been found for other radioisotopes from Sellafield, with ^{137}Cs for example being traced right round the north of Scotland down to the Channel coast of Europe (Murray et al., 1978). Indeed this may be one of the few plus points from such contamination - that it has provided an extremely useful tracer for current sytems and patterns around the coast.

Radioactivity of different origins can be differentiated not only by the radioisotopes released but also by the ratios of the isotopes. For example, $^{239/240}$Pu occurs below detection levels in the natural environment, so its presence is an indication of radioactive contamination.

Figure 3.9   Concentration of $^{239/240}$Pu in *Fucus vesiculosus* with distance from Sellafield. (From Mitchell et al., 1986).

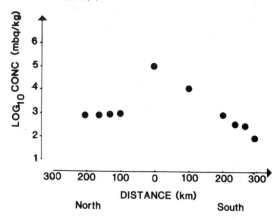

Likewise, the ratio of $^{239/240}$Pu to ^{238}Pu in nuclear test fallout is around 40:1, while the ratio from reprocessing waste disposal approaches 4:1 thus enabling both the identification of the source and also estimation of the degree of contamination (Table 3.11).

Table 3.11. $^{239/240}$Pu:^{238}Pu ratios with distance from Sellafield (see also Figure 3.8; from Mitchell et al., 1986).

Location	Ratio
Ireland, west coast	20:1
Ireland, south coast	10:1
Ireland, Dublin Bay	4.8:1
Ireland, north-east coast	3.8:1

The effects of radioactivity are largely confined to its action as a mutagen and carcinogen, as direct damage through large scale releases (accidental and deliberate) have fortunately been

rare. These effects have only been demonstrated at
the top level i.e. man, and there is no evidence of
radiation damage to the system at a lower level.
However, this lack of evidence is probably due to
the fact that the damage is occurring at a very low
intensity, one which cannot be detected by current
monitoring methods, and it is these methods which we
will go on to consider in the next chapter.

Chapter Four

EFFECTS AND DETECTION

The detection of pollution and its effects are not always straightforward. Going back to the definition of pollution in the previous chapter, it is necessary firstly to prove that there is damage to or some degree of impairment of beneficial use to the system. Secondly there has to be proof that the damage or impairment was caused by some anthropogenic input, and this direct cause/effect link has in many cases been extremely difficult (and long and costly) to prove (see for example the mercury poisoning case at Minimata in Chapter Three). In this chapter we shall look at some of the ways of detecting and monitoring pollution and quantifying its effects by the use of indices where possible.

Indices have many advantages over mere descriptions, notably in their simplification of a complex mass of data and in communication - vital in management. An index selects a component or components from the data mass such that any change in the selected component(s) mirrors the change in the system as a whole. Indices are accepted and in common use elsewhere, for example in the financial world where the Dow Jones Average and the FT100 Share Index are regularly quoted, so the principle is well established. What this chapter will do with selected indices is to set out the rationale upon which they are based, so that their value in a particular circumstance can be judged. It is important to bear in mind at this stage that indices, because they are usually numerical in form, can give the impression of a precision and certainty that would be lacking in a verbal explanation, and so the limitations of the methods and techniques on which they are based should be clearly understood.

It is not the purpose of this volume to detail

the various chemical analytical methods for the
different kinds of contamination, and for this the
reader is directed to works such as Parsons et al.
(1984), or reviews such as Aston (1986). However we
shall briefly consider which part of the system
should be chosen for analysis with the advantages
and disadvantages of each and then go on to examine
in detail the effects that pollution has on the
biological system and how this may be objectively
assessed.

There are three components of the estuarine
system that can be analysed, namely the water
column, the sediment and the organisms themselves,
and their respective merits and demerits are
summarised in Table 4.1.

Table 4.1 Advantages and disadvantages of analysis
of water, sediment and organisms

Parameter	Advantages	Disadvantages
Water	"What's there" Models available Rapid response	High variability in time and space Low levels
Sediment	Integrates Easily sampled High levels	Availability of pollutant?
Organism	Integrates "What's available" High levels	Interspecific differences Survival limits

The great advantage of analysis of the water
column is that the results show the amounts of
contaminant present - important where rapid
detection and quantification of a pollution incident
are required - and the forms or speciation in
circulation. They can be used to assess what the
organisms are likely to be exposed to (but note that
this is not the same as the dose the organism
receives, which depends on how much it can and does
take up) as well as for calculation of the amounts
transported in and out of the system (see for
example the N and P budgets in Chapter Two). The
latter point is important, as there are many models,
both one- and two-dimensional which can be used for
management and prediction of the residence time or
fate of an input. O'Kane (1980) has reviewed the use
of models in estuarine management, and these will be
discussed more fully in Chapter Five.

4.1 INDICES

Pollution monitoring in the marine environment is some way behind that in freshwater, and to a large extent the procedures are based on those which have proved successful in the latter. The index suggested by the National Water Council of the U.K. (see Table 4.2) largely uses the parameters of freshwater eutrophication, since these were the data available (Portmann and Wood, 1985).

Table 4.2   NWC estuarine quality classification scheme (Portmann and Wood, 1985)

Category 1: Chemical quality		Points
Dissolved oxygen exceeds	60%	10
	40%	6
	30%	5
	20%	4
	10%	3
	<10%	0

Category 2: Biological quality
a) Allows the passage of migratory fish (e.g. salmon)	2
b) Supports a resident fish population consistent with habitat	2
c) Supports a benthic community consistent with habitat	2
d) Absence of toxic or tainting substances in the biota	4
Summed Total (Maximum =	10)

Category 3: Aesthetic quality
a) No pollutant inputs or inputs not causing aesthetic pollution	10
b) Inputs causing some aesthetic pollution but no serious impairment of usage	6
c) Inputs affect usage	3
d) Inputs causing widespread public nuisance	0

The scores for the three categories are then added and the total gives the overall estuarine quality. Scores of 30 - 24 indicated good quality, 23 - 16 fair quality, 15 - 9 poor quality and 8 - 0 bad quality. In their survey almost two-thirds of English and Welsh estuaries were good quality and a mere 4% bad quality (Portmann and Wood, 1985). Where this scheme is lacking however is in those cases where there is chemical pollution without nutrient loadings or oxygen demand, leading to an over-optimistic assessment of quality (Wilson and Jeffrey, 1987).

A different approach, but based again on the freshwater experience and data which should be either presently available or easily obtained was put forward by Johnston (1983). This was intended to form a basis for discussion and to suggest the way estuarine indices should develop rather than as a finished and tested article, and so only the framework of this scheme is presented (Table 4.3).

Table 4.3   Outline of an objective environmental quality index (from Johnston, 1983)

1A.   Incidence of disease in man
       Basis: Public Health data
1B.   Troublesome plants and animals
       Basis: biological monitoring
1C.   Amenity etc.
       Basis: water sport activities

2A.   Fish and shellfish quality
       Basis: food standards
2B.   Fish and shellfish disease
       Basis: parasite surveys etc.
2C.   Fish and shellfish habitat
       Basis: %area or value

3A.   Biota contaminated with Annex I substances
       Basis: standards for Annex I substances
3B.   Ecosystem affected
       Basis: biological quality c.f. Trent Biotic Index
3C.   Damage to ecosystem
       Basis: discharge/dumping/extraction records

4A.   Uses of area
       Basis: industrial, tourist or other value/potential
4B.   Biota and use
       Basis: fouling, water extraction costs
4C.   Debris, obstacles, traffic
       Basis: Port Authority reports, costs

Within each category (1A, 1B etc), a score is

given according to whether the effect or use is much worse than normal, worse than normal or normal, and the total score then gives the Index. Johnston (1983) emphasised the flexibility of his scheme, in that the methods of calculating the final index could easily be adjusted or weighted in the light of experience, but concluded himself that a semi-geometric scale, rather than an arithmetic scale, would better meet the demands of the situation as the former was more discriminatory. However, this index remains to be tested, and until then it is impossible to give any opinion on its performance in a real situation.

Returning to Table 4.1, there are basic disadvantages to monitoring the water column. The principal difficulty is the variability of the water column itself in time and space. Because the estuary is a very dynamic system, there are great changes, notably with the ebb and flow of the tides daily and the amount of freshwater input seasonally, in the actual water body at any one fixed point. Since the water body itself is continually changing, the pollutant dose being delivered changes continually also, and consequently a large number of samples spread over the range of environmental conditions is required to ensure an accurate assessment of status. Soulsby et al. (1985a) calculated that, to detect a 25% change in the mean value of certain water column parameters, some hundreds of samples were needed, and that in other cases it was not possible to detect a change of this magnitude at all.

A second drawback to the monitoring of the water column is that pollutants may be present in very low concentrations, making their detection more difficult and also increasing the variability of the results. While there are standard methods to detect such low levels, they do demand a high standard of equipment and technique and commonly involve a concentration step, on an ion-exchange column for example, in the laboratory.

There are samplers which continuously sample the water passing any one spot, and thus obtain an integrated sample over one complete tidal cycle or whatever period might be desired but these are relatively untried as yet (e.g. Fabris et al., 1986). However, a great deal has been done on the use of sediment or organisms as indicators of pollution as these too integrate pollution over time. The pollutant load of sediment or organism reflects its past exposure to that pollutant, and so by sampling them at one point in time it is possible to get an idea of the average pollutant load delivered there.

In addition to their capacity to integrate pollutant loads, both organisms and sediments have the capacity to concentrate pollutants often by several orders of magnitude (commonly ppm as opposed to ppb in water) and this makes detection of those pollutants present in trace amounts that much easier.

Sediment has another advantage as well in that it is relatively easy to collect and practically ubiquitous: there are very few estuarine locations where there is no sediment present.

For these reasons the sediment has frequently been selected for analysis as an indicator of pollution status and can form the basis of the pollution indices such as those proposed by Jeffrey et al. (1985) and Long and Chapman (1985). Implicit in these indices is the concept of an unpolluted or baseline level for pollution, in other words that which one would expect to find in an unpolluted situation. Any levels above this are ascribed to pollution, and the degree of pollution is sometimes expressed as the concentration factor (CF) i.e. as a multiple of the unpolluted baseline level.

However this has one major drawback in that the CF, although it accurately quantifies the amount of pollutant present, and allows intercomparison of the levels of different pollutants in different situations, tells nothing of the effects. To get round this problem, Jeffrey et al. (1985) put forward first approximations to threshold levels for a range of pollutants, the threshold level being that at which damage to the system could be detected, and then scaled their pollution load index (PLI) according to baseline and threshold values. Unfortunately, the data on which to base sediment threshold values is still rather scarce and although there is good agreement between the PLI and other estuarine indices (Wilson and Jeffrey, 1987), Jeffrey et al. (1985) recommended that the threshold values should be periodically reviewed in the light of further information as and when it became available.

This problem highlights the major disadvantage of the analysis of sediment (or water), in that the concentration of the contaminant obtained from the analysis varies greatly according to the techniques used. Chapter Three has already mentioned the techniques of differential extraction for heavy metals as an assessment of their availability in both the sediment and water column, whereby the amount able to affect organisms may be a fraction of the total amount present. In the sediment, the concentration of many contaminants increases as the particle size decreases, since, for heavy metals for example, both the surface area for adsorption and the amount of organic matter for absorption is greater in fine sediments. Not only does this make it difficult to relate pollutant load and effect, it also poses difficulties in the comparison of different sediments, in that a difference in the concentration may be a result of different particle size composition rather than different contaminant inputs. As a result, several workers have advocated the analysis of a standard fraction of the sediment such as the <100μm or <63μm fraction (Bryan et al., 1980; Wilson et al., 1987).

The sediment quality triad proposed by Long and Chapman (1985) put forward a direct solution by incorporating not only chemical data on sediment contamination and faunal data on biological status but also two or more toxicity tests. A variety of tests can be incorporated into the triad, ranging from those in which test organisms are incubated directly with the test sediment to those in which the organisms are exposed to some extract, aqueous or otherwise depending on suspected contamination, of the sediment.

This index has been tried with considerable success in Puget Sound and San Francisco Bay (Long and Chapman, 1985; Chapman et al., 1986), and good agreement was obtained between the three components of the the triad. However, they did note that the chemical data alone could not be relied upon, on a station-by-station basis, for an accurate assessment of the biological effects and that therefore some measure of the latter was essential for accurate prediction.

4.2 BIOLOGICAL INDICATORS

Given the problems of what water or sediment contaminant levels mean in terms of the exposure of organisms, it seems logical to cut out this uncertainty and look at the levels in the organisms themselves. Like the sediment, the organisms act as integrators and concentrators of contamination, and their use as indicators has been advocated on quite a wide scale e.g. the "Mussel Watch" (Goldberg et al., 1978).

Phillips (1980) has reviewed the use of organisms as aquatic indicators and has set out the criteria which an indicator organism should satisfy, namely:

1) The organism should accumulate the pollutant without being killed.
2) The organism should be sedentary to be representative of that study area.
3) The organism should be abundant so that adequate numbers can be obtained for experimental or statistical purposes.
4) The organism should be reasonably long lived, so that more than one year class can be obtained.
5) The organism should be large enough to be able to obtain sufficient tissue for analysis.
6) The organism should be easily sampled in the field.
7) The organism should be hardy enough to survive in the laboratory, to allow defecation before analysis or laboratory studies of pollutant uptake.
8) The organism should tolerate brackish water.
9) There should be a simple correlation between the pollutant concentration in the organism and that of the environment.
10) The correlation between organism and environment should remain constant over the range of habitat and conditions.

Likewise, Phillips (1980) has listed the factors to be taken into account in the design of an organism monitoring programme, and the way in which these factors can introduce unwanted variability to the results. Any such programme should allow for:

a) Seasonal variation, which operates both at the level of pollutant delivery and water chemistry, notably with regard to something like freshwater input, and organism condition and physiology (see also below).
b) Biochemical composition or condition of the organism, especially where certain classes of pollutant are known to have affinity for selected tissues e.g. organochlorines for lipids.
c) Organism size. In many organisms, the smaller individuals have much higher pollutant concentrations in their tissues and the relationship between size and concentration or content can be used to elucidate the routes by which the pollutant

is taken up.
          d) Sex of individuals. While in many species
sex   makes   little   difference   to   pollutant
concentration,   there   are   several   sources   of
variation. For example female gametes (eggs) tend to
have higher lipid than male (sperm) (see notes above
on lipid) and immature individuals do not of course
go through the breeding cycle and have therefore a
different pattern of tissue growth.
          e) Location of sampling site. This is important
in those areas such as estuaries or stratified bays
where there are vertical differences or gradients in
the water body, and the position of the organism
determines its exposure to the pollutant. In
addition to the kind or concentration of the
pollutant, the height on the shore of an intertidal
organism will determine the length of time for which
the pollutant is presented; the higher on the shore,
the shorter the time it will actually be exposed to
the pollutant in the water.
          f) Behaviour. Many species have the ability to
detect pollutants or harmful or noxious substances
in the water column and to take appropriate action,
by moving away or by closing the shell, to avoid it.

          The great disadvantage of the use of organisms
is the fact that they may have quite specific
requirements as to habitat or environmental
tolerances. Wilson and Jeffrey (1987) have pointed
out that these requirements mean not only a
restriction of the geographical range over which
their use is possible, but also that organisms do
not survive in areas of high pollution or even
perhaps high environmental (natural) stress, and
that therefore their use as indicators is limited by
their tolerances. Such species like the mussel as
have been suggested for monitoring, tend to be more
tolerant than most (see Phillips' (1980) criteria 7)
and 8) above) and perhaps therefore their reaction
is not typical of the biological system as a whole.
However Bayne (1985) recommends the use of
eurytopic, tolerant species for pollution assessment
at species level, as their tolerance is often based
on a flexibility of response which can then be
measured. At community level, it may be preferable
to concentrate on those species with other
strategies, and this point will be discussed a
little later.
          Bryan (e.g. Bryan et al., 1980; 1985) has
examined in detail the use of several organisms as
indicators, and has concluded that while a

particular organism may be a good indicator for one
or two pollutants, they are by no means universal
indicators. As general indicators of dissolved
metals, the algae (*Fucus*, *Enteromorpha*) were best,
although contamination of the fronds by sediment
could lead to problems. The polychaetes *Nereis* and
Nephthys were good indicators for copper, but they
were weak accumulators and tended to actively
regulate some metals such as zinc. The bivalves,
including the widely-used *M. edulis*, were generally
good accumulators, although here again there were
signs that some metals were regulated, but the
deposit feeders were held to be useful indicators of
sediment contamination (Bryan et al., 1985). Clearly
then, the choice of indicator organism needs to be
made with careful reference to the known or
suspected pollutant.

Even using organisms as indicators, it is
difficult to correlate body pollutant loads with
damage or impairment to the individual. Many
organisms have mechanisms for deactivating and
storing many types of contaminant within their
tissues where they are virtually harmless. The
pollutant only takes effect when they are released,
either through the organism itself metabolising
tissue or through the digestion of the tissue when
the organism is eaten. In this way organisms can
accumulate body burdens well in excess of those
noted at death in laboratory experiments, and in
fact the exposure to pollution may stimulate their
ability to survive even higher levels (Bryan and
Hummerstone, 1971).

4.3 DOSE/RESPONSE RELATIONSHIP

The relationship between dose and response
varies according to the contaminant(s) and
organism(s) involved with both interspecific and
individual differences in tolerance. The general
shape of such a relationship is sigmoid and the
concept is expressed diagrammatically in Figure 4.1.
Most toxicity studies have concentrated on the top
end, that is on concentrations causing death,
usually in rather a short time (commonly 24 or 96
hours), and this is commonly expressed as the LD_{50},
the dose required to kill 50% of the population.
However, it should again be emphasised that this is
more properly expressed as the LC_{50}, in other words
the concentration required to kill 50% of the
population, as "dose", as has been mentioned above,

implies what has actually been taken up by the organism. Since exposure to contaminants is unlikely to be restricted to 24 or 96 hours in nature, except perhaps in special cases where a one-off spill or incident occurs (and here it might be argued that the proper exposure time would be one tidal cycle i.e. about 12.5 hours), an asymptotic LC_{50}, that is where the plot of LC_{50} versus time levels out, is clearly preferable. However, there are certain practical difficulties such as the problem of feeding and contaminant transfer in and out in food and faeces involved with the longer time scale, and it is probably easier to estimate this statistically with probit analysis (Bliss, 1935).

Figure 4.1 Dose/response relationship.

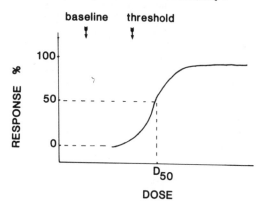

Even so, extrapolation downward to those concentrations that would be "safe" is extremely uncertain, and the use of a safety factor of 0.1 or 0.05 of the LD_{50} as has been suggested (NAS/NAE, 1973) for discharges does not seem to hold as well in the marine environment as it does in freshwater.

Note also in Figure 4.1 that at the bottom end there is commonly a residual effect even in the absence of any contamination, and this can best be thought of as corresponding to something like the background cancer rate in a population. Where the background rate is high, the detection of low level pollution effects is that much more difficult.

Just as the organism's previous exposure can modify its lethal tolerance to a contaminant, it can also modify its sub-lethal tolerances (within the absolute limits of tolerance). This is illustrated in the following table (Table 4.4), which shows how the upper limit for the temperature tolerance of the bivalve *Tellina tenuis* can vary according to previous exposure and the parameter selected for performance.

The highest temperature tolerated is that for the shortest lethal test (24 hour LD_{50}), next is that for the longer LD_{50}, and temperatures for the sub-lethal effect, burrowing is lower still, and Ansell et al. (1980) in fact considered that a good idea of the 96h LD_{50} could be gained from the 24h BT . All these must be borne in mind when making the decision on the amount of thermal pollution that could be discharged into a particular location, and in this example it is noteworthy that although the Mediterranean animals have higher tolerances, the difference in tolerance (about 1 - 2° C) is much less than the environmental difference (some 10° C). The maximum difference between environmental temperatures and LD_{50} was about 23° C (Scottish animals, winter) and the minimum was a mere 8° C (Mediterranean animals, summer) (Ansell et al., 1980).

Table 4.4 Temperature limits for *T. tenuis* (Ansell et al., 1980; Wilson, 1976)

Population	Parameter	Temperature limit (°C)
Scotland	LD_{50} (24h)	31.25
	LD_{50} (96h)	30.20*
	Burrowing (BT_{50})	28.75*
Mediterranean	LD_{50} (24h)	32.80
	LD_{50} (96h)	30.75*
	Burrowing (BT_{50})	30.45*

*summer/winter mean value

Temperature and salinity can significantly affect the tolerances of organisms to contaminants. In their review, McLusky et al. (1986) concluded that in general the susceptibilty of estuarine animals to metals increased as salinity decreased and temperature increased and that toxicity values determined under fixed (or single) temperature/ salinity regimes may be inappropriate for evaluation of the effect such concentrations may have in typical estuarine situations where both parameters may be expected to show considerable short- and long-term variation.

Equally difficult is relating the delivery of pollutant in laboratory studies to delivery in the field, again tied up with the problem of LD_{50} and LC_{50}. Most laboratory studies of metal toxicity for example have used doses of ionic metal, but as shown in Figures 3.6 and 3.7, there may be only a small fraction of the metal available in the field in this form. Short-term laboratory experiments naturally preclude realistic assessment of long-term effects, and it is also difficult to gauge the effect of pollutant delivery via the food chain, say, as opposed to direct uptake from the water. It is possible to scale up the experiments by using mesocosms or macrocosms to enclose and manipulate some segment of the natural environment, but experience with them, though promising, is limited (Davies and Gamble, 1979).

Again the answer seems obvious: to find out if there is damage to the system it is necessary to look at the system itself. Within the system there are a myriad of functions or processes which can be investigated, and these can be broadly grouped into a sort of heirarchy as in Figure 4.2.

Figure 4.2   Heirarchy of organisational levels of investigation.

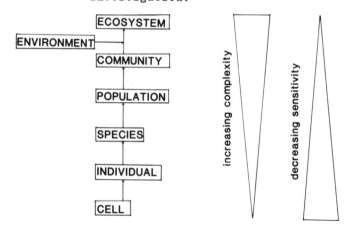

As a general rule, the more complex the system, the more difficult it is to distinguish and attribute changes within it (see for examples the comments on monitoring the water column above), but this has to be reconciled with the fact that it is an overall assessment of the system that is desired, and that it is clearly impossible to assess each component separately! For this reason, "sensitivity" in Figure 4.2 is more a reflection of the ability (or inability) to detect pollution change than of the pollution effect itself.

Just as for the indices described earlier in this chapter, biologically-based detection of pollution change or damage has to be quantified against the situation that would have existed in the absence of such pollution, in other words a baseline. To a large extent it is the reliability of this baseline that determines the accuracy of detection of change: the better and less variable the baseline, the easier is detection.

4.4 COMMUNITY RESPONSES

With field techniques, the existing status quo has to be compared either with that before the advent of pollution or with a similar situation elsewhere. In most cases, the former is just not available, or if available is incomplete, and the latter is hampered by the fact that there will almost certainly be differences, geographical, hydrographical, sedimentological or whatever, between the two locations and these differences in themselves will introduce variation.

This problem has greatly hampered the development of effective ways of quantifying pollution damage at the top i.e. community level (Figure 4.2). Under gross pollution, all life disappears but on the gradient between this level of contamination and normal conditions different species respond in different ways. Pearson and Rosenberg (1978) have summarised these in the classic species/abundance/biomass (SAB) curves (Figure 4.3) in which the characteristic community response to moderate pollution is a combination of a few species which do very well under such conditions and of a biomass which can be higher than in an unpolluted situation.

There have been several attempts to use this property in pollution monitoring, for instance by using those few abundant species as pollution indicators, such that their presence would indicate a certain level of contamination as is the case with freshwater indicator species. Unfortunately, it

quickly became evident that these species were not
limited to polluted locations, but could be found
under a variety of more or less unpolluted
conditions.

Figure 4.3   Community response with distance from a
              pollution source: after a) Pearson and
              Rosenberg (1978); b) Leppakowski
              (1977); and c) Wilson and Jeffrey
              (1987).

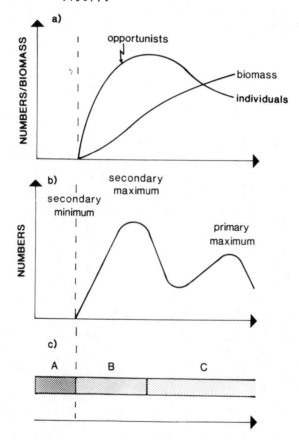

Pearson and Rosenberg (1978) in their review of
pollution effects considered them rather as
"opportunistic" (of which more later) or
"transgressive" species than as indicators, and
included a list of them from the literature. What is
evident from their list is that many of these

species are those typical of estuarine conditions, and that therefore organisms must find in the estuarine environment conditions not dissimilar to those in a polluted situation.

Returning to the SAB curves for a moment, Leppakowski (1977) proposed that the different communities with distance from a pollution source could be assigned to different categories which he termed primary and secondary maxima and primary and secondary minima (Figure 4.3b). The areas assigned to each category could then be mapped in the field and the pollution status calculated from the formula

$$BPI = 10 - (5x + 4y + 2(z - a/2))/a$$

where a = whole water area, x = area of the secondary minimum, y = area of the secondary maximum, and z = area of the primary maximum.

A similar idea was put forward independently (Tomlinson et al., 1980; Jeffrey et al., 1985), but this time reducing and simplifying the idea to the three zones abiotic, opportunistic and stable (Figure 4.3c). These three zones corresponded to the areas A) without life, B) with low species numbers but high biomass, and C) with normal species and biomass under Pearson and Rosenberg's (1978) scheme. Again the zones were mapped in the field and the proportion of the estuary assigned to each category was put into the formula for the Biological Quality Index (BQI) (Jeffrey et al., 1985) such that

$$BQI = antilog_{10} (C - A)$$

where C = the proportion of the estuary assigned to the "stable" category and A = the proportion assigned to the category "abiotic". A totally abiotic estuary (A = 1) would score 0.1 and a totally unpolluted estuary (C = 1) would score 10. In practice, BQI values ranging from 0.1 to 9.95 have been recorded from a variety of European estuaries, with good agreement between the BQI and other methods of quality assessment (Wilson and Jeffrey, 1987). This approach comes close to that recommended by Johnston (1983), whose own index outline was presented in Table 4.3, in that it depends on the rough characteristics of the community and not the detailed enumeration of the species. As such, it is intended to serve only as a crude measure of environmental quality for management and not for scientific purposes.

An alternative approach is to look at the species richness or heterogeneity, using indices such as the Shannon-Weiner Index (H') which is widely used in freshwater pollution assessment. Under polluted conditions, the numbers of a few species increase dramatically (see Figure 4.3a) and the organisation of the community changes. The degree of change can be quantified by these indices, sometimes referred to collectively if rather misleadingly as "dominance indices", and a variety of these have been reviewed by Gray and Pearson (1982). Variations on this approach include the index of evenness ($J = H'/H_{max}$, where H' = diversity and $H_{max} = \log_2 s$ with s = number of species), and Gray (1981) recommended that both diversity and evenness be given. The reason is that an increase in the Shannon-Weiner diversity statistic can be due solely to an increase in the number of species found (since $H'_{max} = \log_2 s$) and not to any increase in the actual heterogeneity of the system.

Gray (1976) himself used the Shannon-Weiner in polluted estuaries in the north-east of England, but Wilson (1983) noted that low Shannon-Weiner values (usually taken to be indicative of pollution) were obtained from a variety of sites, estuarine and open coast, high and low shore, polluted and unpolluted in Dublin Bay, Ireland. The conclusion is that these types of index are not suited to the estuarine situation or any other in which there are naturally few species (hence low diversity) or in which some species may be extraordinarily successful (hence low heterogeneity).

Shaw et al. (1983) have proposed that a simple graphic method, the rank species abundance (RSA) curve (Figure 4.4), in which the species found are ranked in order of abundance and then rank is plotted against abundance, could be used to measure dominance, and hence pollution. As an added advantage, they suggested that the rarer species could be omitted from the plots, considerably easing the problems of species identification.

Figure 4.4 shows the RSA plots from four situations, two showing the shape of curve to be expected from a stressed and an unstressed community respectively (Shaw et al., 1983) and two from estuarine situations: Bull Island, Dublin which is slightly polluted and Bannow Bay, SE Ireland, which is unpolluted (Magennis, 1987b). As with other dominance indices, it can be seen that although the curves appear to be very sensitive to even a slight degree of pollution, some stress does seem to be indicated in an unpolluted estuarine community. However, it does seem that the task of identification is eased somewhat, in that over 95% of the total numbers of individuals from Bull Island were accounted for by just the first five ranked species (Figure 4.4).

Figure 4.4   RSA curves showing theoretical stressed
and unstressed curves (broken lines)
with plots from a slightly polluted
(open circles) and an unpolluted (closed
circles) estuarine situation.

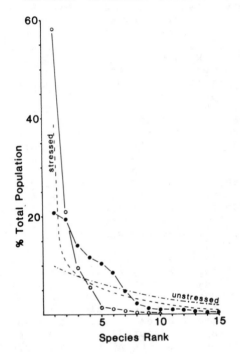

    More   recently   Platt   and   Lambshead   (1985)
suggested that deviation from a neutral model of
diversity could be used to detect disturbance or
stress (including pollution).   The   V-statistic is
calculated from the formula
$$V = (H' - EH')/\sigma EH'$$
where H' = observed diversity, EH' = theoretical
diversity and σEH' = standard deviation of EH'. When
V = 0   the sample is adjudged to have come from a
"neutral" assemblage,   V > 0 indicate excess
equitability and V < 0 (i.e. negative values) excess
dominance. Applying this to a variety of different
data sets, though not unfortunately a specific
estuarine set, they conclude that the V-statistic

could be used to detect disturbance but that the direction of change (positive or negative) might be dependent on the amount of disturbance.

The estuarine environment is highly dynamic, and the physical and chemical fluctuations can put a lot of pressure on the organisms which live there. Gray (1979) considered this problem and proposed that there are two kinds of pressure, namely stress which operates at a relatively low intensity but is more or less constant and disturbance which is high intensity but intermittent. The strategies which organisms can adopt are summarised below (Table 4.5).

Table 4.5   Possible pollution strategies (after Gray, 1979)

		Disturbance	
		High	Low
Stress	High	None	Tolerant
	Low	Reproductive (r-strategy)	Competitive (K-strategy)

Of course the choice is not as stark as presented in Table 4.5, and the Reproductive-Competitive-Tolerant strategies should rather be viewed as continua such that species will align themselves according to the proportions of the different elements. Estuarine species align themselves strongly with the "Reproductive-Tolerant" axis, while in comparison stenohaline truly marine species are highly competitive (Figure 4.5).

The reproductive strategy is to produce large numbers of offspring to take advantage of any vacant niches that may be available - hence the name "opportunists" - and this may be as successful in estuaries after a particularly violent incident such as a storm flood as after pollution. One of the species first suggested as a marine pollution indicator, *Capitella capitata*, is a classic r-strategist and it has recently been shown that rather than being a single species, *C. capitata* is actually a complex of six sibling species within which there is a succession from most opportunistic downwards in response to a pollution incident (Grassle and Grassle, 1974). In other words, the species (*sensu lato*) *C. capitata* responds in exactly the same way as the community as a whole.

Finally, we can consider using the species-richness approach developed from the log-normal distribution proposed first by Preston (1948) and developed for use in pollution monitoring by Gray (Gray, 1979; Gray and Mirza, 1979; Gray and Pearson, 1982).

Figure 4.5 Alignment of species on Reproductive-Competitive-Tolerant gradients (after Gray, 1981).

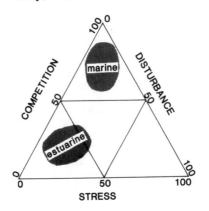

The numbers of individuals in each species are assigned to geometric classes, and the cumulative number of species or the number of species is plotted against the geometric classes (see Figure 4.6). Under moderately polluted conditions, some species increase in abundance but the numbers of the rarer species are not affected, with the result that the community does not conform to the log-normal distribution, and the plot deviates from a straight line to assume a "broken stick" arrangement. Under heavier pollution, the community re-establishes conformity with a log-normal distribution albeit different from the original, and the slope of the line, although the plot itself may be straight, is considerably depressed indicating a high degree of dominance by a few extremely numerous species while the others are few in number (Gray and Mirza, 1979).

The shape of the curve is characteristic for each type of community, and several workers (Andrews and Richard, 1980; Gulliksen et al., 1980; Gray and Pearson, 1982) have used it successfully to demonstrate pollution effects in a variety of (mainly non-estuarine) situations. A useful comparison of the log-normal and other techniques including the Shannon-Weiner index has been given by Andrews (1984), in which it clearly demonstrates the recovery of the Thames estuary during the 1970's.

The experience of Rygg (1986) with heavy metal pollution was somewhat different, in that the "broken stick" arrangement was not found under any intensity of contamination and all communities conformed to the log-normal distribution. The difference under chemical contamination was ascribed to the fact that this causes a reduction in all species, rare or common, unlike the situation under organic enrichment where some species' numbers may be considerably increased.

Despite these successes, Magennis (1987b) concluded that its use to assess estuarine pollution might be limited, in that even though it was still possible to distinguish between the unpolluted and the slightly-polluted sites, a stress response was still detected at the unpolluted location (Figure 4.6).

Figure 4.6   Log-normal curves at an unpolluted (■) and a slightly polluted (●) estuarine site (after Magennis, 1987b).

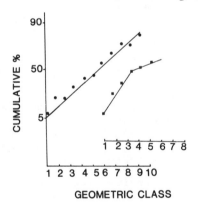

Other criticisms have come from Shaw et al.

(1983) and Platt and Lambshead (1985) in which it is compared much to its disadvantage with the neutral model (above).

The conclusions from these works is therefore that conditions in estuaries may evoke much the same community changes as pollution, and in consequence it may prove difficult in practice to separate the two. Wilson and Jeffrey (1987) have in fact suggested that there are parts of estuaries which are naturally abiotic or opportunistic, and that the boundaries of these areas may change with time and environmental fluctuations. Given this, then contamination may be better visualised as an additional stress whose effect should be considered in the context of these natural fluctuations, and this is a point which we will return to later.

4.5 POPULATION RESPONSES

Given then the problems imposed by the fact that there are very few species in estuaries, is it possible to assess pollution status from the performance of one of these, that is at the level of the population? How well a population is doing can be judged from how well it is growing, that is from the production, P per unit area (usually expressed as g/m^2), or per unit of starting biomass, B, as P:B ratio. This approach has advantages, not least because the species selected may be of commercial value and the data collected could serve for stock assessment or exploitation potential as well as putting a direct economic cost on the effect of pollution. Commercial species have also usually been extensively investigated, and there will be a considerable bank of other work with which to make the comparison in addition often to records of catches or landings going back many years to give some historical perspective. Even if a commercial species is not available or indicated, there are other estuarine species whose populations, for a variety of reasons, have been well studied.

An example of such a species would be the bivalve *Macoma balthica*, for which data exists from many estuaries in Europe and North America and to which more is being added under programmes such as the EEC COST 647. Elliott and McLusky (1985) considered the use of productivity as a measure of detecting sub-lethal stress to be particularly useful for a number of reasons and illustrated it with a case study of *M. balthica* from the Forth estuary in Scotland. Within well defined limits, it was possible to show that populations in polluted areas had lower production and productivity (as P:B ratio), but there was considerable variation even within the one geographic area.

Figure 4.7 gives an example of growth rates and P:B ratios from three estuaries: the Ythan and the Forth in Scotland and Bull Island in Ireland (Chambers and Milne, 1975b; Elliott and Mclusky, 1985; and Magennis, 1987b respectively).

Figure 4.7   Growth rates and P:B ratios in three populations of *M. balthica*, Forth (●) Ythan (○) and Bull Island (■).

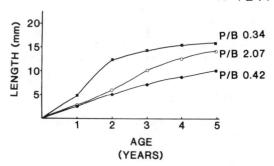

The population from the Forth estuary, which is the most polluted, has both the lowest growth rate and a very low P:B ratio. The Bull Island population has the lowest P:B ratio, but Magennis (1987b) considered this to be a result of the population structure, which was dominated by older individuals, rather than of any pollution effect.

Elliott and Mclusky (1985) recognised this difficulty and recommended limiting the comparison to populations with similar age spans, while Magennis (1987b) has suggested that the technique may only be applicable when specific year-classes are compared.

These difficulties would appear to confine for the present this technique to a few specific situations where either the production itself is a desired datum e.g. for commercial purposes or where there is a sufficient body of growth and productivity data to allow for the variables mentioned.

4.6 INDIVIDUAL RESPONSES

The logical next step down from the population or the age-class is the individual itself, and here there are a plethora of methods that have been used to assess pollution status and effects.

Perhaps the simplest of the individual indices, though often used in conjunction with more sophisticated techniques, are the body condition indices, which require only dissection and weighing of the specimen, or perhaps a simple chemical extraction. GESAMP (1980) highlighted their advantages of low cost and direct ecological significance, but did point out that they can be variable, insensitive and slow to respond to environmental change. They are often used in shellfish research, particularly in commercial studies where meat yield and condition determine the financial return. A selected list is displayed in Table 4.6.

Table 4.6   Body condition indices

> flesh wet weight/dry weight
> flesh weight/shell weight
> flesh weight/shell volume
> shell inorganic ($CaCO_3$)/organic (periostracum)
> % tissue lipid
> liver weight/body weight
> gonad weight/body weight

Because of their ease and simplicity they are ideally suited to study of pollution gradients with samples at intervals from the suspected source of contamination. Figure 4.8 shows an example of this in which the body condition indices of the mussel *M. edulis* (flesh/shell and periostracum/$CaCO_3$) and the winkle *Littorina rudis* (flesh/shell) change with the tissue copper content which in turn was linked to distance from the copper source (Wilson and McMahon, 1981).

The basis of these indices is the same as the population productivity assessment described above. The greater the pollution, the less the growth and hence the smaller the amount of flesh. The lipid content of tissues can also be used as lipid is one of the ways in which the body stores energy; under stress less is laid down or more of the reserves have to be utilised. The exception is the liver weight/body weight ratio which has been used with some success in fish (Heidinger and Crawford, 1977), where the ratio increases as the size of the liver grows to cope with the extra detoxification.

The great drawback to all these indices, with the exception of the shell inorganic/organic ratio, is that the ratios change with season and also with the age of the organism. To be comparable therefore, all samples must be of similarly-sized specimens taken at the same time of year, and if possible checked as to reproductive status (male, female, immature) and state (unripe, ripe, spent). Within these limits, condition indices have proved useful and sensitive, and they show good agreement with other more sophisticated or specialised techniques (GESAMP, 1980).

Other workers have used respiration (or photosynthesis for primary producers) as an indication of the degree to which the metabolic performance of the individual has been affected by pollution, and in certain cases a correlation has been found between the contaminant load and the respiration rate (Figure 4.8).

Figure 4.8   Body condition indices (ratios) and Cu concentration (µg/g) in *M. edulis* and *L. rudis*. From Wilson and McMahon (1979).

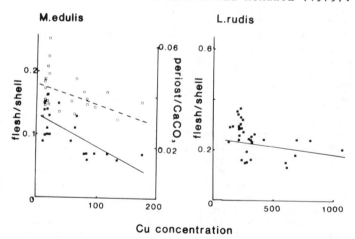

Cu concentration

However, Wilson and McMahon (1979) did point out that the effect may be to a certain extent species-specific, and while an absolute difference may be detected, it is difficult to quantify the relationship because of the many variables that can affect respiration rate.

A variation on this is the ratio of oxygen consumed to nitrogen excreted (commonly referred to as the O:N ratio). Low values (<30) signify that a high proportion of the total energy requirements of the individual are being met by the catabolism of protein relative to carbohydrate and lipid which usually comprise the body's energy reserves (Bayne and Scullard, 1977).

As with the population productivity, the condition of the individual reflects its past exposure to environmental factors, - here it is worth noting that natural heat stress for example or poor feeding conditions can affect condition as pollution does - and they therefore give a good integrated picture over the life-span of the organism. A more instantaneous measure using the same principle has been developed (Bayne et al., 1979) from the individual energy budget

$$P = A - (R + U)$$

where A = assimilation, that is the difference between what is consumed (C) and that which goes straight through and is lost as faeces (F), R = respiration, U = energy lost as excreta and P = production (somatic and reproductive). Production, P depends on the amount of food available, the efficiency with which the individual can extract energy from it (i.e. (C - F) or A) and the demands made for routine maintenance of tissue metabolism, R and excretion, U. What is left after these routine losses is available for new growth or reproduction, hence the name given to this measure, "scope for growth".

With this technique, Bayne et al. (1979) have shown that the scope for growth of mussels (M. edulis) from a polluted site amounted to just 2.7 kJ for a standard animal of 1g dry flesh weight, compared to 32 kJ for an animal from an unpolluted site. As can be seen from these data, the mussels from the polluted site were barely able to maintain condition over the year, and it is possible to get a negative scope for growth where the animals would be losing condition. The validity of these measurements has been confirmed by the summing of scope for growth values calculated monthly to predict total individual growth over a number of years and obtaining a close match between the predicted sizes and those actually observed in the field (Bayne et al., 1979). Scope for growth therefore gives an open-ended assessment of exactly how well an organism is doing under a certain set of circumstances, and does in addition offer the possibility of manipulating the experimental set-up. Again this has been demonstrated by Bayne et al. (1979) who transplanted mussels to polluted stations in Narragansett Bay, Rhode Island and obtained, after thirty to forty days exposure, a gradient in the scope for growth at each site to match that of the pollution in the Bay.

As with condition, scope for growth is affected by natural environmental variables as well as pollution, and can be severely reduced by unusually warm weather or parasitic infection (Bayne et al., 1979). Also like a condition index, because it is measured at the level of the individual, it will vary from individual to individual, and it has been suggested that the performance is genetically controlled with heterozygotes being "fitter" than homozygotes (e.g. Garton and Stickle, 1985).

Stebbing (1976, 1979) proposed the use of a colonial hydroid, *Campanularia flexuosa* as a bioassay organism, as this can be cultured from a single individual and therefore all members used in a test will be genetically identical. In addition this test is designed to be rapid (14 days), to incorporate replicates and to lend itself to manipulation of experimental conditions in that the assay is performed in the laboratory and the results expressed against unpolluted control water values. The hydroid assay is extremely sensitive, and offers a range of parameters, namely growth rate, stolon curving (deformity of growth) and gonozoid frequency (number of reproductive individuals) each of which responds to different sub-lethal levels of pollution. In fact this extreme sensitivity may be a drawback to the technique, as Magennis (1987a) has suggested that the organisms may be affected by the air quality via the aeration of the chambers! While *C. flexuosa* is not an estuarine organism and hence its use is restricted to situations where the salinity does not fall below 30‰ , there are other species which do better at lower salinities and to which the technique can equally well be adapted (Stebbing, 1981a).

The advantage that the hydroid assay, like most invertebrate techniques, has over the use of fish is of scale and speed. Boelens (1987) has reviewed the use of fish in water pollution studies and has concluded that, despite the fact that the presence and survival of fish in a particular location is

often taken as a sign of unpolluted conditions (see
for example the criteria for the NWC index in Table
4.2), their selection as monitoring indicators is
often difficult to justify. In particular he drew
attention to the difficulty of selecting the "most
sensitive species of local importance" as
recommended by the US Environmental Protection
Agency (1973) without prior experience or data on
relative sensitivity, and pointed out that the
majority of work was with freshwater conditions and
species. The difficulty is partly related to the
preference, in the European context, of species
which can be cultured in pollution-free conditions
in the laboratory or other controlled conditions. In
North America there appears to be a greater
aceptance of the use of wild stocks, provided that
they come from an unpolluted environment, and one of
the most widely used species is the sheepshead
minnow, *Cyprinodon variegatus*. Nevertheless, the use
of different species does pose problems in the
intercomparison of results and their extrapolation
to different situations, and there is always the
doubt that the test species have been chosen for
other reasons (ease of collection, hardiness or
ability to survive in the laboratory - see Phillips'
(1980) criteria for indicator species) than that
they are the "most sensitive species of local
importance".

The sensitivity of some of the sub-lethal
responses of fish, responses such as changes in
heart beat or opercular rhythym or avoidance, can be
put to good effect in effluent monitoring. By
putting such a test system "on-line" in the effluent
stream at a constant dilution or range of dilutions,
an almost instantaneous response can be seen to a
change in effluent chemistry and the appropriate
steps can be taken to prevent this reaching the
receiving waters. Such a system can also be used to
measure the toxicity of a presumptive effluent, and
used as any other sub-lethal test to assess the
degree of dilution that might be required or the
effect once in the system.

Field measurements of sub-lethal pollution
effects on fish have concentrated largely on the
area of pathobiology. While this initially appeared
to offer a promising avenue of investigation
(Bengtsson, 1979), some authorities consider that
much of the evidence remains equivocal or at best
circumstantial (Mulcahy et al., 1987), and the
suggestion is that pollution acts indirectly by
stress or immunosuppression on the organism's

defences except in the few cases where known mutagens or carcinogens directly affect gene expression. None the less, it is important particularly in the public perception, as evidence like skeletal deformities, skin lesions or dermal papillomas is easily seen and the incidence of such or the incidence of reporting of such is increasing. These sorts of investigations are also important because of the commercial importance of fish and shellfish, and many argue that the weight of evidence justifies more control over discharges on these grounds alone. A particularly strong case was made recently in the case of the North Sea, where there have been many reports of disease and deformity (Dethlefsen, 1984) and this was a powerful argument in the hands of the conservation lobby for curtailment of certain inputs.

There are many variations on this approach, and one of the most common is to use the eggs or larvae of invertebrates in place of fish. Apart from the obvious handling advantages, there is a considerable body of evidence to suggest that these stages are much more sensitive to environmental and contaminant stress than the adults, and by examining the effects of pollution at the most sensitive level, it is possible to get a more accurate idea of the effects in the field. For these reasons, laboratory tests routinely use sea urchin eggs; correlations have also been shown in the field between pollution status and larval abnormalitities in other species such as the winkle *Littorina 'saxatilis'* (Dixon and Pollard, 1985).

4.7 CELLULAR AND SUB-CELLULAR RESPONSES

That organisms can adapt to their environment is well known (see the example of .   , Table 4.4 above), and the ability to adapt extends, albeit perhaps to a lesser degree, to pollution. This phenomenon, known as homeostasis, can mask the response and for this reason many workers have advocated cytochemical tests in preference to the physiological or "whole animal" approach. Cytological or biochemical tests are therefore likely to be more sensitive, and in addition, early warning of possible deleterious effects further up the system (see figure 4.1) may result from indications of damage at the cellular level. These tests have also been investigated to find out if it is possible to identify a unique cellular response to a specific pollutant or group of pollutants and give them a diagnostic character lacking in the other approaches.

Although the subject has been reviewed by Moore et al. (1986), the conclusions are that there is still a dearth of information as to the precise mechanisms of toxicity, and much of the knowledge available on mammalian toxicology does not translate to invertebrates. As a result, the field itself is still developing, and few of the suggested molecular or cellular indices have been widely tested. Table 4.7 shows a selection of these indices to show the range and diversity currently being put forward.

Table 4.7 Molecular and cellular indices: see text for explanations

Index	Reference
Amino acid ratio	Jeffries (1972)
Adenylate energy charge (AEC)	Ivanovici (1980)
Lysosyme latency	Moore et al.(1986)
Digestive cell epithelial thickness	Lowe et al. (1981)
Reproductive cell volume	Lowe and Pipe (1985)
Sister chromatid exchange (SCE)	Dixon and Clarke (1982)
Lectin binding	Simkiss and Schmidt (1985)
Enzyme systems:	
a) Inhibition	
Phosphofructokinase	Blackstock (1980)
Pyruvate kinase	Blackstock (1980)
b) Stimulation	
NADPH-tetrazolium reductase	Bayne et al. (1979)
Aryl hydrocarbon hydroxylase	Lee et al. (1981)

The change in amino acid ratio was first noted by Jeffries (1972), who found that a whole range of stress responses in the clam *Mercenaria mercenaria* could be summarised by examination of the molar ratio of taurine to glycine. In normal, unstressed animals, the ratio is less than 3, values between 3 and 5 indicated chronic stress and values in excess of 5 denoted acutely stressed individuals. The major drawback of this technique in an estuarine situation is that bivalves such as *Mercenaria* use amino acids to osmoregulate. As the external salinity falls, the concentration of free amino acids, particularly taurine, increases. The amino acid molecules are large and therefore do not easily pass out of the cell, and in this way the animal is able to maintain internal osmotic pressure.

The adenylate energy charge (AEC) is the cellular equivalent of the scope for growth in that it measures the amounts of energy present and available to the cell. The AEC is calculated from the formula

$$AEC = (ATP + 1/2ADP)/(ATP + ADP + AMP)$$

where ATP, ADP and AMP are the measured concentrations of the adenylate pool, namely adenosine-5-triphosphate, -diphosphate and -monophosphate respectively. The theoretical range of the AEC is from 0.0 to 1.0, but it has been suggested that the upper limit is normally between 0.8 and 0.9, and the lower limit for viability and survival is around 0.5 (Chapman et al., 1971). Ivanovici (1980) pointed out that measurement of the AEC required relatively simple field collection techniques and the results showed a precision and a marked lack of variability when compared to other parameters such as the ATP/ADP ratio. However, McElroy (1985) found the ATP/ADP ratio more sensitive than the AEC in detecting pollution effects in the polychaete *N. virens* and noted also that the AEC itself was lower (below 0.75) than expected in the control groups. This difference in interspecific and interphyletic AEC response was one of the limitations in the technique listed by Ivanovici (1980), who emphasised that the species concerned may have to be selected carefully and that abnormally high or low AEC values have been noted under certain circumstances both in whole organisms and in selected tissues.

The lysosyme latency assay is one of a suite of indices developed by Bayne and his co-workers (see e.g. Bayne et al., 1979; Moore et al., 1986). When scope for growth in an individual is negative, the energy demands must be met from within the animal itself, that is by autophagy of its own tissues by the lysosomal enzymes. Normally held latent by the lysosomal membrane, interference with or disruption of the membrane results in the release of these enzymes into the cell, and the degree of membrane instability is measured cytochemically by the length of time taken for maximum activity (measured by intensity of staining). The length of this labilisation period denotes the degree of latency, i.e. how near the cell is to self destruction. Good correlation has been obtained between this index and a variety of others, and also between the latency period and the degree of pollutant exposure, but it should be borne in mind that it is a generalised response to stress and cannot be used to identify specific stressors (Bayne et al., 1979). Using this technique, Magennis (1987b) found that pollution differences between two locations were obscured by the stress response generated by unusually warm conditions and concluded that the applications of this index, despite its simplicity and sensitivity might be limited.

The following index, namely the digestive cell epithelial thickness (Table 4.7) is likewise a development of the cellular disturbance approach but based on disturbance of cell structure rather than function. A decrease in the thickness of the epithelia of the digestive cells with hydrocarbons and other toxic substances has been noted in *M. edulis* and *M. mercenaria* respectively (Lowe et al., 1981; Tripp et al., 1984). The mechanism behind these changes is not yet fully understood, but it is thought to be linked to the lysosome stress response, and the thinning of the epithelia is a result of increased autophagy of the cell (Moore et al., 1986).

The lysosomes are also thought to be agents in the decrease in reproductive cell volume which has been noted in several types of reproductive cell in *M. edulis* under hydrocarbon contamination (Lowe and Pipe, 1985). Again it is increased autophagy by lysosomal enzymes that are thought to mediate the response, and again the link with the energy available (scope for growth) is evident.

Considered solely as pollution indices, it is obvious that the latter two have been relatively little tried, and until more evidence is available, it may be better to consider their principal value as a better basis for the understanding of the mechanism behind pollution effects. Given also their link with the lysozyme assay, which has been much more extensively used, both as regards the situations and the species involved, it would seem premature to use these two without evidence of some greater sensitivity or precision, as this link would also seem to indicate that they too will be a non-specific stress response.

Many contaminants are carcinogenic or mutagenic, and the effects of genetic damage may result in an increase in the numbers or the proportion of the population with deformities, as in fish pathology mentioned above. This damage can also be detected at the chromosome level, either by identifying directly structural abnormalities in the chromosome or by studying sister chromatid exchange (SCE), based on the switching of labelled arm segments within chromosomes (Dixon, 1983; Dixon and Clarke, 1982). The latter has been shown to be particularly sensitive to the action of mutagenic agents (Dixon and Clarke, 1982), although Simkiss and Schmidt (1985) have criticised it on the grounds that it is technically demanding and may not show sub-lethal effects. In its stead, they proposed (Simkiss and Schmidt, 1985) a lectin binding technique to identify the different membrane glycoprotein and glycolipid composition which they considered the most sensitive indicators of changes in genetic make-up. However, despite some success with initial trials of the technique in a test clone, this has still to be more fully tried in a field situation.

The enzyme assays (Table 4.7) can be conveniently divided into two groups: those in which the enzyme is impaired by pollution, in effect a sort of biochemical condition index, and those in which the activity is induced or enhanced by the pollutant.

In the first category many enzyme systems have been tested and Bayne et al. (1985) provide a list of no fewer than 32 that have been tested in a variety of marine organisms. The list (Bayne et al., 1985) shows clearly that not only in many cases have the results been somewhat variable even within the one taxon, but also that different taxa can give completely different results with the same enzyme/stressor combination.

Similarly, different systems exhibit different degrees of sensitivity within the chosen organism. The selected examples (Table 4.7) are from the work of Blackstock (1980) with various enzyme systems in the polychaete *Glycera alba* . Phosphofructokinase showed the most marked decline in activity under polluted conditions, pyruvate kinase displayed a similar pattern, although the decline was rather less and the results with ∝-glycerophosphatase and malate dehydrogenase showed no difference. Thus it can be seen that just as the different body condition indices can respond to different degrees under pollution stress, so can the different enzymes display different sensitivity. As Blackstock (1980) himself pointed out, these assays are performed *in vitro*, and there is no guarantee that the effects noted in the laboratory can be extrapolated directly to the whole organism response in the field. None the less, phosphofructokinase and pyruvate kinase are key enzymes in glycolysis, and diminished activity will clearly decrease energy yield. In this context of extrapolating laboratory results to field consequences, the enzyme selected for assay must be the rate controlling enzyme in the pathway or cycle; if not, then a change in enzyme activity need not be reflected in a change in the overall pathway flux.

The whole field of enzyme response to environmental change and the effect of contaminants has been recently reviewed by Blackstock (1984), particularly in regard to the sources of variability in and the range and types of the response. In this, as in the other methods of this type, more work needs to be done before it can be adopted on a wider scale for routine pollution assessment.

The second group of enzymes (Table 4.7), often referred to collectively as the mixed function oxidases, are those responsible for the breakdown of a variety of foreign compounds including hydrocarbons and organochlorines. These enzymes are specifically involved with one compound or one class of compounds such as aryl hyrocarbon hydroxylase (AHH) which as its name suggests is involved in the metabolism of hydrocarbons. Activity of this enzyme has been shown to be enhanced under hydrocarbon contamination both in laboratory and field situations (Lee et al., 1981), although other evidence is somewhat variable and equivocal (McElroy, 1985).

Also involved is the cytochrome P-450 system (including linked enzymes such as NADPH-tetrazolium reductase) and the whole process of breakdown of foreign compounds by the monooxygenases, which are also stimulated - when exposed to hydrocarbon contamination for example (Moore et al., 1986). While this may be valuable as a generalised pollution stress assay, it lacks the specificity of something like AHH.

For reasons of specificity therefore, this line of investigation looks particularly promising, as the ability to detect not only effect but also cause is one which is missing from almost all pollution assays.

However promising the prospects, the enzyme stimulation does pose one particular problem - how does one relate enhanced performance with the concept of pollution, remembering the definition which demanded "deleterious effects"? In fact the dilemma is not confined to this particular set of circumstances alone. It has already been shown how biomass or productivity can be elevated under certain conditions of organic and nutrient enrichment by the removal of the normal checks and balances on these processes. Different again is the stimulation which has also been demonstrated in response to low levels of persistent pollutants like metals, and Stebbing (1981b) who termed the process "hormesis", considered this to be a consequence of the over-compensation of the homeostatic mechanism. It is possible to visualise this homeostasis in action at each stage of the integration of the system shown in Figure 4.2 such that at each level there is a mechanism by which adaptation to the changes may be made (see e.g. Blackstock (1984) on compensation and adaptation at enzyme level).If compensation is not possible because the effect is too strong, then it is passed on up to the next level (see e.g. Capuzzo, 1985).

Lowe and Pipe (1987) give an example of how a sytem might cope with pollution. They noted that exposure of mussels to hydrocarbons resulted in a reduction in the reserves of stored energy and an increase in gamete degeneration and resorption. As a result of the latter, energy resources within the organism were re-allocated, such that although the next generation was affected (fewer gametes), the individual itself was better able to tolerate the hydrocarbon insult.

Although completely different in character, the fact that there exist ways in which some functions of the system can be enhanced demonstrates the difficulties of detecting subtle pollution-induced changes. Unless such responses lead to some impairment of the system, then the level of whatever it is causing the response cannot be termed pollution, but we can use them as early warnings of deleterious effects to come if the contamination were to increase.

4.8 OVERVIEW

GESAMP (1980) attempted a ranking of 37 different biological variables that could be selected for marine pollution monitoring. The majority of the ecological variables (diversity, abundance, biomass etc.) were ranked 1, that is highly recommended for immediate use in all regions, while the biochemical variables (such as the enzyme assays and AEC) were ranked 2 or 3, that is in general needing more development of the technique.

One of the recurring themes throughout this

chapter is the difficulty of extrapolating
laboratory results to the field. This is partly due
to the problems of contaminant presentation or
availability which has been discussed already, but
the great drawback is the sheer complexity of the
situation in the field. Most laboratory tests or
assays present substances singly under carefully
controlled conditions; to do otherwise would be to
introduce too many variables into the experiments
for any meaningful answers. Unfortunately
contaminants in the field rarely come singly, and
in most polluted situations it is possible to find
elevated levels of a whole range of contaminants.

In this situation, the presence of more than one
pollutant may have no effect or may have greatly
increased effect (synergism) or even greatly reduce
the effect (antagonism). At this point it is worth
considering also whether the stress of the
estuarine environment acts in a similar manner, and
various workers have shown that in general, the
sensivity of estuarine species to pollutants is
greatly modified by the various abiotic factors
(e.g. Theede, 1980).

What is clear is that the approach advocated by
those trying to devise estuarine quality indices (as
opposed to indices designed to quantify deviation of
some sub-component of the system from normal
values), is that the index must take into account
chemical and biological data and what is actually
happening in the field. In contrast to the
freshwater situation, which is comparatively well
studied and less complicated, estuarine (and marine)
indices have proved difficult to set up. Conversion
of freshwater techniques has not proved entirely
satisfactory, either in status evaluation or indeed
in the assumptions behind them (diversity and
pollution, safety factors for toxic discharge etc.).
However, it is also obvious that a great deal has
still to be done before any one method for estuarine
quality assessment can be adopted in the way
freshwater assessment has pioneered, and it is
probably due to the current absence of an accepted
scheme that there are so many alternatives being
developed and tested.

In the management context there are two
questions that will be asked. Will a particular
assay work in a specific situation, i.e. will it
give a true picture of the pollution effects? Can it
be performed given the resources available
(expertise, facilities, finance)?

To a certain extent the answer to the first

question depends on what is required: damage to some commercial interest such as fishing may indicate a much narrower suite of indices than would be indicated by an assessment of overall pollution status. The location also must be considered. While it has just been emphasised that the freshwater approach has not been satisfactorily adapted to the estuarine system, it should also be admitted that the majority of the assays mentioned above have been designed and tested more in marine than estuarine conditions, and as a result, their use may be restricted to the seaward end. There are few, for example the PLI and BQI of Jeffrey et al. (1985), which have been designed specifically to meet the difficulties of the estuarine situation, and many respond to the stresses imposed by the system as they would to pollution stress (see comments on various indices above).

Similarly there are few which do not demand some level of scientific expertise or technical facilities. Although Bayne (Bayne et al., 1979) considered that the kinds of tests developed by his group could be widely used by others in that the measurement of the indices was simple enough for most laboratories, he did concede that the interpretation of the data might require some expertise, and the few attempts that have been made to use indices from other laboratories have not been wholely without problems (e.g. Magennis, 1987b).

In any event, where an index depends on a certain level of expertise or familiarity for interpretation, it is difficult to achieve complete objectivity, a point that was emphasised by Jeffrey et al. (1985) in their index justification. However, since the very definition of pollution is itself to a certain extent subjective, perhaps complete objectivity is too stringent a criterion! Biological variability is such that it is impossible to lay down hard and fast rules on what constitutes "deleterious effects" or "damage", but as more and more information becomes available, perhaps it will soon be possible to make a better informed judgement as to how best to detect it, and to make the selection of a pollution index on scientific rather than largely practical grounds.

Chapter Five

MANAGEMENT

5.1 GOALS

The estuarine system cannot cope with the conflicting demands being made on it without proper management, and the aim of this chapter is to set out the objectives and practices of good management.

Many of the management practices and customs have grown up or been arrived at in an way through the development of the estuary. Each problem has been tackled as it has arisen, and there has been little in the way of anticipatory action to prevent conflicts arising. As a result, a great deal of energy is spent in the defence of the entrenched *status quo* in which one particular usage of the estuary effectively precludes any change.

Since it is clearly impossible to please all interests all of the time, the goal of management must be to optimise the use of the estuary, in other words to please as many of the interests as much of the time as possible. The management plan must also be flexible, to accommodate the changes that will inevitably occur in attitudes and priorities in time, and to be able to take advantage of any advances in knowledge, for instance in techniques or effects of waste disposal, that may arise. This means that the current status and future priorities must be kept under review and updated and the management team must keep themselves informed of these changes, either through their own monitoring programmes or from outside sources. Finally any management system must take into account the public perception of its actions and include communication as one of the prime aims. Insufficient public and political backing for the management objectives and practices will raise serious obstacles to their success, and a great deal of time and energy can be dissipated in pointless conflict which could well have been avoided.

The major part of the problem lies in reconciling not only seemingly irreconcilable uses (e.g. industry and conservation) but also the different units by which each is evaluated. While a direct economic value can be put on industrial usage, it is much more difficult to express conservation value in any objective terms, and this has already been touched on in Chapter One. While value judgements may be difficult, an environmental impact assessment (EIA) along the lines of that shown in Chapter One (Figure 1.8), can identify the areas of conflict, and should be an essential first step in any development plan.

Nevertheless, since conservation and wildlife protection are becomingly increasingly advocated as major aims of estuarine management there are important implications as to the management system. Does it for instance imply that the eventual goal of the management plan should be a first class, totally unpolluted estuary?

Given the present state of many estuaries such an objective would be totally out of the question at least in the short term, and perhaps even in the long term, even if the will (and finance) were there. The logical next step would therefore seem to be an acceptance that some estuaries or parts of estuaries will never be first class, and some secondary quality objective, tacitly accepting some degree of contamination, is then adopted. While this may be unacceptable to some, it is obviously a more pragmatic approach, but still leaves the question as to what would be acceptable as a quality objective in that particular situation. Here some sort of cost/benefit analysis is called for, but again there is the problem of evaluation.

Some workers have suggested that management objectives could accept that some estuaries need not be improved; those with gross pollution could be reserved for industrial or other use (such as shipping) for which water quality is unimportant. Since these estuaries are so polluted, the argument runs, additional loading will have no additional effect and there is no question of their having any residual value for other uses like fishing or amenity. They are therefore free to accept additional loadings from new industries and the effort and money spared from attempting a clean-up can be better used in less polluted situations where a real benefit will accrue. Much the same argument

can be used to justify the choice of a containing
dumping site for toxic material, particularly where
the practice has been going on for a number of
years, the argument being that the site is so bad
that it cannot get any worse and it is better to put
the waste there than risk polluting somewhere else.

The fallacy of this argument is that pollution
can rarely be contained as effectively as its
proponents may think. To accept an area of gross
pollution means also accepting an (undefined) area
with a gradient of pollution damage centred around
it (see for example Figure 4.3). Although things
cannot get worse within the grossly polluted area,
the area affected round it will depend on the amount
going in and the characteristics of the accepting
system, and obviously, the bigger the input, the
bigger the total area affected. Apart from the
purely practical aspects of this problem, it means
that the consequences of decisions and actions
cannot always be confined to those responsible, and
this can lead to friction between localities
(states, countries or whatever) who may have
different views on environmental quality. Some of
these problems will be discussed later in the
context of national and international standards and
legislation.

5.2 UNIFORM EMISSION STANDARDS AND BEST AVAILABLE TECHNOLOGY

There are two ways to control the amounts going
into the system. The easist way is to set standards
for the discharge, standards which must be met by
all operators. This is an industrial process
orientated approach, and it is easier because each
individual concern can be given a set of standards
according to the type of process and contaminant
involved, data which are easily available. It has
also the advantage that once limits have been
decided for a particular type of industry, then they
can be applied to all discharges of that type
without further investigation. This is known as the
uniform emission standards approach (UES).

For very toxic substances, it may be necessary,
where a discharge cannot be prevented, to insist on
the maximum reduction in the discharge through use
of the best available technology (BAT), regardless
of the costs of such treatment. There have been many
calls for the application of BAT to the radioactive
discharges from the nuclear reprocessing plant at

Sellafield, for example. The same approach may be indicated where a very valuable resource has to be protected or where there are obvious signs that sustainable use of the system is under threat. In the absence of such considerations, and where treatment costs are a factor a compromise can be reached between cost and degree of treatment - the best practical means available (BPMA). As with all compromises this solution may be unsatisfactory to all parties, since the eventual emission standards set usually depend on the weight given to the economic argument. With old plant in particular, modernisation or tightening of standards may be difficult and expensive, and, in the absence of concrete evidence of pollution, this may be used as an argument against any improvement.

5.3 ENVIRONMENTAL QUALITY OBJECTIVE

The disadvantage of the UES approach is that of course a large discharge into a small system will result in a much lower environmental quality than a small discharge into a large system, and the alternative is to look at the problem from the other end and set an environmental quality objective (EQO) by means of environmental quality standards (EQSs) for the range of parameters. This presents a much more difficult problem since the eventual environmental quality depends on the amount and type or toxicity of contaminant going in, the size of the receiving system and its capacity to cope with the contaminant load.

Perhaps the most important of these factors is the toxicity of the contaminant as discussed in Chapter Three. For management purposes, the aim is to set a maximum allowable concentration (MAC) or maximum acceptable toxicant concentration (MATC) which must not be exceeded in the receiving water. The MAC or MATC for individual pollutants can be set directly from toxicity tests or sub-lethal assays as described in Chapter Four, or an alternative approach, that of critical path analysis (CPA) can be adopted.

The latter is particularly useful either where specific target organisms (usually man) can be identified or where the contaminant in question is known or is suspected to be capable of bioaccumulation. The CPA approach is most widely used in the field of radiological protection, where the groups most likely to be at risk can be identified along with the pathways by which the radiation is delivered (see e.g. Hunt, 1987). By back calculation from the internationally agreed safe dose that can be tolerated, through the amounts consumed and the ability of the food items to bioaccumulate, a level for concentration in the water column can be set. A hypothetical example, which assumes the critical pathway is through food consumption, is shown in Table 5.1 below.

Table 5.1 Hypothetical example of CPA to derive a water quality standard

Allowable dose per person		1,000 units
Background		100
	Margin	900
Food Consumption per person		10kg
=> Maximum permitted concentration in food		= 0.09 units/g
BCF from water to food		10,000
=> Maximum concentration in water		= 0.009 units/l

The calculation starts with an agreed dose or body burden per individual; from this is subtracted the background from the environment to leave the "safe" margin that may come from anthropogenic sources. As the entry route is via food, the amount ingested per unit time (say one day or one year) is then measured and this calculation gives the maximum permitted level for the food. Then the bioconcentration factor (BCF) from water, into which the substance is being discharged, to the food is calculated, and this gives the maximum permitted concentration that could be allowed in the water column – the first approximation to the EQS for that substance.

In addition a safety factor can be built in that would reduce the safe level in food as well as in the receiving water. In the example in Table 5.1, a safety factor of 0.1 would result in a water EQS concentration of 0.0009 and a food concentration of 0.009, so that the critical group receives a dose equivalent to just under twice the background level (190 units in total rather than 1,000).

In practice the calculations will rarely be as simple as shown above. In many cases the desired objective is the protection of entire biological communities rather than of a single (well-studied) species, the pathways are by no means well defined

148

and the contaminants in question are discharged as a variable mixture of substances. However, the CPA approach does offer a way out of the rather arbitrary definition of MAC or MATC that has prevailed until recently, and with regular monitoring and feedback, the accuracy of the derived standard can be kept under review.

Of course the toxicity of the effluent can be measured directly by exposing test organisms to the discharge. The toxicity can be expressed in toxic units (TU) = $100/LC_{50}$, and WHO/UNEP (1982) suggested the following formula for calculating the toxicity of an urban sewage discharge

$$Toxicity = 0.74 + MBAS/8 + (NH_3-N)/46$$

where MBAS and NH_3-N are the concentrations (in ppm) of surface-active substances and ammonia nitrogen respectively. Primary treated effluent contains approximately 2.21 TU equivalent to an LC_{50} of 45%, chemical precipitation and sedimentation can reduce this to 1.5 TU and biological treatment to around 0.5 TU (WHO/UNEP, 1982). The toxicity of an effluent is considerably increased by chlorination, and benthic diversity and vitality have been adversely affected at a concentration of 0.04 TU (WHO/UNEP, 1982).

The next step is to assess the capacity of the receiving waters, and this lends itself well to mathematical modelling. These at their most basic can be simple mass-balance calculations based on a black box approach (see for example N and P balances, Figures 2.10 and 2.12). In a physically well-defined system like a lagoon, where water exchange with the open sea is restricted, losses to the system are small and therefore the capacity as a waste receiver is limited. In more dynamic systems such as estuaries or open bays with strong tidal action or longshore currents, advective transport can remove a major part of the loading, always provided that these are not major sources of input as well (see for example the hydrocarbon budget in Figure 3.5).

In this situation, one difficulty is fixing the boundary conditions for the modelling exercise, although O'Kane (1980) considered that most estuarine situations could be adequately modelled by a simple one-dimensional model. Two dimensional models are indicated only where there are marked vertical salinity gradients or where length and width of the estuary are comparable, and in general water quality models should be kept as simple as possible in terms of their hydrodynamic base (O'Kane, 1980).

Most water quality models currently being used in estuarine management are based on conservative substances, or on those whose stoichiometric relationships or transformations within the water column can be fairly well predicted (e.g. O'Kane, 1980; Riddle, 1985). This has led to a reliance on water chemistry, and biologically mediated transformations or sediment interactions are usually treated empirically (O'Kane, 1980). Many biologists distrust such a simplistic approach and in return many modellers and engineers will not accept judgements based on parameters which cannot be rigorously defined and tested.

More complex models are available, and a whole new series are being developed in which the reaction of the biological component of the system can be modelled as well as the physical and chemical e.g. the GEMBASE model (Radford, 1981). However, it must be remembered that the more complex the model, the greater the time needed for development and verification and consequently the greater the costs.

At the very least the model should have some predictive capability, that is by manipulating the variables in the model it will show what will happen under the complete range of input and environmental conditions. In many situations also, even a simple model based on the minimum of data can provide valuable information on the system response, highlighting areas or components of greatest sensivity, and this can be incorporated into any subsequent sampling or monitoring programme.

The object behind this calculation of assimilative capacity is to ensure that neither is the limit exceeded nor is it under-used. For economic reasons, it would be wasteful either to insist on over-stringent and perhaps unattainable discharge criteria or to ban development altogether and planning strategies for this will be discussed later. It must be understood however that the acceptance of this approach is an acceptance of some degree of contamination of the system, albeit at below the damage threshold.

The removal of any contaminant discharged into the system depends on three main processes:-

1) Physical removal or transport out of the system, either by water currents or through loss to the atmosphere;

2) Removal within the system, such that it becomes unavailable, for example by burial in the sediment;

3) Breakdown or transformation of the contaminant, either by biological or other means such as photo-oxidation.

The first two mechanisms are by far the most important, in purely practical terms of how waste is at present discharged (see for example the hydrocarbon budget in Figure 3.5). They do depend however on the secondary receiver (the adjacent water body, sediment) being able to accept the input: what happens when the sediment for example becomes saturated is that no more contaminant will be removed from the water column. Similarly with water transport; what is removed on one tide may be largely brought back by the next, and in those situations where there are water/sediment fluxes which vary with salinity or oxidation state, the net movement may be difficult to predict. In the Elbe estuary, net movement of metals is upstream, although net movement of water is of course out to sea down the estuary (Lichtfuss and Brummer, 1977).

Fortunately, estuaries are in general highly dynamic, so that dilution and dispersion of discharges are easily achieved, and this physical process greatly increases the reactive volume for breakdown and transformation. Care must be taken however that removal from one system does not just result in accumulation in another less dynamic system. Such a problem arose in Loch Eil in Scotland where pulp mill effluent discharged into the turbulent entrance to the loch accumulated in the bottom of the loch basin - at some distance from the input, it should be noted - smothering and choking the benthos (Pearson, 1982).

GESAMP (1986) considered that most physicochemical processes such as hydrolysis or photo-degradation as well as biological degradation could be adequately described by a first-order reaction of the form

$$[P]_t = [P]_o \cdot e^{-Kt}$$

where P_o is the initial pollutant concentration, P_t is the concentration at time t and K is the rate constant.

The same equation can be used for dilution, and a model calculated for removal by the integration of

the component removal pathways. Curran and Milne
(1985) provide an example of this for bacterial
removal from the Clyde estuary in Scotland, where
the three causes of disappearance considered were
deposition, physico-chemical parameters such as
salinity, and finally sunlight, each of which gave a
different disappearance rate corresponding to T_{90}
values of 185, 55 and 2.1 hours respectively. A
wider range of examples with a variety of different
kinds of contaminant has been given in the WHO/UNEP
(1982) publication "Waste discharge into the marine
environment".

Once the rate of removal has been established a
limit can be set on the amount of contaminant that
can be discharged, by back-calculation from the MAC
or MATC which sets the final quality standard
desired in the receiving water.

The great drawback to the widespread use of the
assimilative capacity approach is the amount of work
involved. Apart from the difficulties at the
discharge end, that is what's there, how much and
how it behaves, the solution is site specific. In
other words, only a small part of the knowledge
gained in one situation will be of use in the next,
as the actual capacity of the receiving system
depends on its individual and often unique
characteristics such as the size and shape of the
estuary or bay.

5.4 DISCHARGE LOCATION

Such considerations were of course far from the
mind of those who put in the first disposal systems,
and the original discharges were designed simply to
take the waste to the nearest watercourse,
irrespective of subsequent use. Present practice on
the siting of outfalls or sludge dumping grounds
demands a rather more rational approach, usually
with a view to obtaining maximum dilution of the
discharge. The criteria for the selection of such
sites have been set out in some detail in GESAMP
(1982), including the restrictions on the types of
wastes that can be discharged under international
Conventions which will be discussed later.

GESAMP (1982) recognised that the marine
environment does have a capacity to absorb waste and
that the aim of site selection should be to minimise
the impact on the environment. In recent years, the
emphasis has shifted from judgement of impact on
visual or aesthetic grounds to one in which

accumulation of persistent pollutants has forced the consideration of much longer-term effects on the functioning of the system as a whole. If the assimilative capacity of the system is exceeded, then either there arises a direct danger to human health because the contaminant is not being neutralised by whatever mechanism(s) quickly enough, or the system itself is degraded, and that degradation decreases its capacity as a waste receiver and the deterioration process is accelerated.

It is difficult to generalise on the siting of discharge points for effluent or sludge, since each situation has its individual characteristsics which must be considered, and different sets of priorities which will decide the final outcome. However, there are some points that can be made.

The main point, and one which is often ignored, is that the site selection should be on the basis of the receiving system not the discharging i.e. because it is suitable for the reception of the waste rather than because it is adjacent to or convenient for its land-based origin! There is clearly no point in going to the time and trouble (and expense) of calculating the assimilative capacity of a system as a whole and then abusing it by concentrating the loadings in one randomly-chosen location.

Arising from this is the need for an adequate survey not only of the immediate area, but also of any adjacent areas that might be affected. Although dilution and dispersion may be fairly considered as part of the assimilative mechanism of the system, care must be taken to ensure that the introduced substances are in fact neutralised and not merely transported, possibly to be accumulated elsewhere (see the Loch Eil example mentioned above). As was shown in Chapter Three, the fate of contaminants is not random, but is controlled by their various physical, chemical and biological properties. To take the example of a contaminant which binds strongly to sediment particles, it may be rapidly transported away (bound or unbound) from its point of discharge but may accumulate in high concentrations in any areas where water movement is low and sediment settles out of suspension.

5.4.1 Sediment Criteria

A totally different set of criteria would apply

in the case of a containing site, in which the object is to prevent the dispersion of a toxic waste such as dredge spoil from a polluted harbour. The US EPA has set limits (the Jensen criteria) beyond which a sediment should be considered toxic, and a similar approach has been adopted by the State of Connecticut which classifies dredge spoils into Class I - nontoxic, Class II - moderately toxic but suitable for dumping, and Class III - potentially hazardous, which must be proved non-toxic by bioassay before they can be ocean dumped. Selected parameters are shown below (Table 5.2) along with the baseline and threshold values proposed by Jeffrey et al. (1985) for their Pollution Load Index (see Chapter Four).

Table 5.2   Jensen criteria, Connecticut State limits and PLI baseline (B) and threshold (T) values (data from Kester et al. (1982) and Jeffrey et al. (1985))

Parameter	Jensen	Connecticut			B	T
		I	II	III		
oxygen demand	50mg/g	-	-	-	-	-
organic content	-	-	-	-	1%	5%
Total N(mg/g)	1.0	-	-	-	0.4	2.5
Total P(ug/g)	-	-	-	-	150	500
Hg (ug/g)	1.0	0.5	0.5-1.5	1.5	0.05	1.5
Pb (ug/g)	50	100	100-200	200	10	100
Zn (ug/g)	50	200	200-400	400	20	100
Cu (ug/g)	-	200	200-400	400	5.0	50
Cd (ug/g)	-	5	5-10	10	0.5	1.5
oil/grease (mg/g)	1.5	3	3-10	10	-	-
hydrocarbons (ug/g)	-	-	-	-	50	600

The object here would be to select a site where water movement and transport away is minimal and to isolate the pollutants from the system by containerisation or by capping with clean sediment. The proponents of such a method of waste disposal would argue that any release into the environment would be so slow and on such a small scale that the system could easily cope, and it has been shown for example that the capping sediment starts to develop a normal community very quickly, within three months (Germano, 1983). The opponents on the other hand argue that such predictions have not been rigorously tested, and that in the event of them being found wanting by monitoring after the event how is the situation to be put right? Capped sites may always be re-capped every few years, an expensive and in itself a disrupting process, but very toxic waste such as radioactive waste in containers would be extremely difficult to find and recover, not to mention the problems of handling it after retrieval.

Fortunately, the disposal of waste so toxic as to require containerisation is rare, and under proper classification (see Table 5.2 above) should not be part of the remit of the estuarine manager. In any event the sites chosen for disposal are likely to be well away from coasts and estuaries, although this does not prevent their occasional appearance on the beach after storm or wreck.

Currently alternative means of disposal are being sought for many wastes and dredge spoils such as those in New York Harbour. In this case (Kamlet, 1983) disposal options under discussion included 1) subaqueous borrow pit; 2) upland disposal in exhausted sand and gravel pits; 3) ocean disposal (including capping); 4) containment island with a dyke inside the harbour; 5) selected "special case" sites with strict limits as to amounts.

In any event, site investigation should be as thorough as possible and extend into neighbouring systems to ensure that the problem is not merely transferred (in either direction)! Quetin and de Rouville (1986) estimated that the costs of a full site investigation amounted to between 0.5% and 3% of the total costs of the disposal system (depending on size), and that proper design and planning at this stage would result in savings in excess of the design costs.

5.5 ENVIRONMENTAL QUALITY STANDARDS

As in all matters, the economic factors have to be considered, hence the suggestion mentioned before that it is not cost-effective to consider improvement at all in some cases. Nicholas (1985) used the phrase 'controlled deterioration' to describe the process by which discharges are licensed until the EQO is reached, and pointed out that this resulted in progressively more stringent conditions being set for new inputs as the spare capacity was used up. Nicholas (1985) proposed rather that the process should be viewed as a progressive series of steps in which discharges could be allowed and improvements made until the EQO was met (Figure 5.1). In this scheme, spare capacity could be allocated until the EQO was reached and any

Figure 5.1   Long-term planning of EQO in three hypothetical estuaries (Nicholas, 1985).

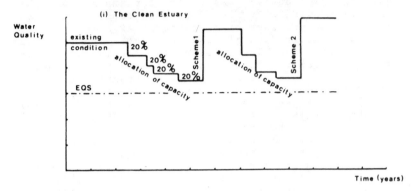

(i) The Clean Estuary

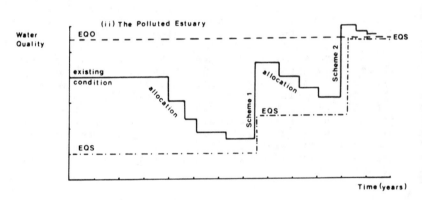

(ii) The Polluted Estuary

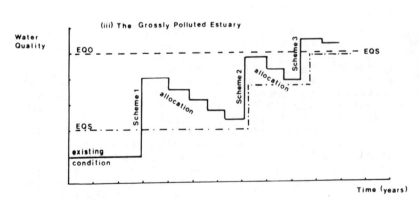

(iii) The Grossly Polluted Estuary

further inputs would have to depend on improvements
being made. However, these inputs need not
necessarily wait for the additional capacity,
provided that such arrangements were on line for the
future.

From a biological point of view, there are
problems with this approach, in that deterioration
below the EQO may cause damage out of all proportion
to the change in the chemical quality (i.e. the
relationship is not linear - see Figure 4.1) and the
subsequent recovery of the system is by no means
assured, yet the EQO is usually set by reference to
biological parameters. These include the passage of
migratory fish, the maintenance of a normal benthic
community and the levels of contaminants that can be
accumulated in the flesh of those organisms consumed
directly by man. For example the water quality
classification employed by the District of Columbia,
USA does so by reference to human health and fish
life (Table 5.3).

Table 5.3   Water quality categories, District of
             Columbia (EPA, 1978)

Category	Designated Use
A	Water contact recreation
B	Wading
C	Fish and wildlife propagation
D	Recreational boating
E	Maintenance of fish life
F	Industrial water supply

While such EQOs or categories are easily
formulated, there have been great difficulties in
translating biological objectives into chemical
standards and considerable research (e.g. Gardiner
and Mance, 1984) has gone into the establishment of
appropriate EQSs. An example of the standards for
two estuarine locations, the Humber in England and
Chesapeake Bay, which properly includes a number of
estuaries within its system, in the USA, is shown in
Table 5.4.

In both locations, the primary objective was to
protect the existing use of the estuary along with
the protection of important fisheries. The two sets
of standards are very similar, but note however that
while the Chesapeake standards stipulate that for
copper and lead LC_{50} values should be established,
they employ also the concept of safety margins. The
major difference is the absence of bacterial quality

standards in the Humber which has no areas
designated for shellfish or bathing. This point will
be returned to later. Chesapeake on the other hand
is famed for its shellfish (see Chapter One) and in
consequence the limits for contamination are
strictly set down, in terms not only of the median
value (Table 5.4) but also of the permitted
variation and the maxima according to the technique
used for assessment (EPA, 1978).

Table 5.4   Selected water quality standards proposed
             for the Humber estuary, UK (Sayers, 1986)
             and for Chesapeake Bay, Virginia, USA
             (EPA, 1978): metals as dissolved, ug/l
             unless specified; ns - indicates not
             stated

Parameter	Humber	Chesapeake
DO	>55%	>4.0 mg/l
Coliforms	none set	<70/100ml (median)
pH	6.0 - 8.5	6.0 - 8.5
Hg	<0.5	<0.1
Cd	<5.0	<5.0
Cr	<15.0	<100.0
Cu++	<5.0	0.1 96h LC_{50}
Pb	<25.0	0.01 96h LC_{50}
PCBs	ns	none permitted

The reason of course for the insistence on
bacterial standards in Chesapeake is to protect an
important commercial activity, and economic (or
pragmatic, depending on the viewpoint) considerations can greatly affect the EQSs set. The
regulations in most of the States in the USA for
example start by affirming that no deterioration of
existing quality will be permitted, and then adds,
unless it can be justified on economic grounds (EPA,
1978).
    Mackay et al. (1986) give an excellent
illustration of the number and the complexity of the
arguments to be considered in setting standards. In
addition to the straight economic argument, it was
possible for two groups of scientists (management
and industry) working from essentially the same data
bank to come to quite widely differing conclusions
about the amounts that could safely be discharged.
In particular discussion centred around the size of
the mixing zone. The mixing zone is that immediately
around the discharge point in which the standards
for the whole system are exceeded since dilution and

dispersion are not instantaneous processes. Within this zone there will be a gradient or plume of contaminant concentration from the source and hence by definition the limits of the mixing zone denote what would be an acceptable area of pollution.

As in most situations, a safety margin is built into a licensing system like this, usually by allocating less than the total assimilative capacity, with the limit set around 70%. The greater the uncertainty in the calculations, the more conservative the allocation of capacity. Spare capacity also allows the possibilty of plant malfunction causing an overload on the system. In this case a single large discharge will call for a greater reserve than many small ones, which are unlikely all to go wrong at the same time.

The point to emphasise here in management terms is that it is often extremely difficult to remedy a situation where the original discharge conditions have not been sufficiently stringent; quite apart from the reluctance of the discharger to change an operation which, to their perception, is going well, it can be technically difficult and expensive to radically change a treatment plant which may have been built into the manufacturing process at construction.

5.5.1 Economics

In recent years there has been a shift from the attitude that resources are there for the taking to one of "shared ownership", that is everyone has a right to share them. Anyone who wishes to appropriate that right must compensate society at large and hence the notion of the "polluter pays" which is now incorporated in much of the legislation.

How much the polluter pays varies greatly. Some countries might demand high treatment standards, others, particularly those wishing to attract industry and investment, may demand little or nothing. While the latter attitude is perfectly justifiable if there is adequate environmental capacity - take for example the situation in Figure 5.1a - it can cause problems if not properly controlled, and considerable concern has been expressed by many who see the process as one in which the developed nations are exporting their pollution to the Third World.

There are various formulae by which treatment

can be costed, and the final figure depends largely on the amount of capital investment to be made not only in the plant itself, but also in the sewerage system to gather and then dispose of it. For domestic wastes, a rough indication of the costs, based on a fairly large installation is shown in Table 5.5.

Table 5.5 Costs (UK £ per head as of 1977) of treatment of domestic sewage based on a population of 45,000 (WRC, 1977)

Treatment	None	Primary	Secondary	Tertiary
Receiving Water	Sea	Estuary	River	Lake
Standards (ppm)	-	150/200	30/20	10/10/10
Costs:				
Civil eng.	1.31	9.47	17.49	22.93
Mech. eng.	0.91	2.00	5.42	12.71
Total	2.22	11.47	22.91	35.64
Operational	0.01	0.02	0.40	0.51

The standards in Table 5.5 (suspended solids/BOD/ammoniacal nitrogen) are the rough targets for each stage of treatment, and in many instances can be bettered in practice. Note, however that the Table does not imply that these standards are acceptable for each of the four receiving waters - rather they are an indication of what was considered acceptable in 1977 - but they simply become progressively stricter as the size of the receiving system diminishes. From the WRC (1977) costings, therefore, sea or estuarine disposal is a very cost-effective option. The above costings (Table 5.5) assume no untoward problems such as the necessity for a long outfall (add roughly 15 per head per kilometre of outfall) or having to transport the sludge long distances for disposal.

The most widely used formula to calculate the length of outfall required is the Pomeroy formula (Pomeroy, 1959)

$$N80 = kQ^2/yx^2$$

where N80 is the standard required, Q the flow in cubic metres/day, y the water depth (m) and x the outfall length (m). WHO/UNEP (1982) quote as an example a required coliform standard of 100/100 ml, and assuming a bottom slope of 10%, the required

outfall length becomes
$$x => n^{2/3}$$
where n = population served. In the example above
(Table 5.5) n = 45,000, and the outfall length would
be just under 1.25 km, adding around £19 to the
capital cost.

Deering and Gray (1987) have reviewed the
charging sytems in Europe and the USA, and have
shown that the charges for the disposal of
industrial wastes vary from authority to authority,
even within the same country. All base the charge on
the amount of waste and its oxygen demand, but some
include parameters such as nitrogen load (Holland),
suspended solids (UK, USA), chlorine demand
(Pittsburg, USA) or make a special charge for metals
(Holland) while some in the UK charge for handling.
The formulae used to calculate the charge vary also,
with the result that average treatment costs in the
USA were around one-fifth of those in Holland and
around one-third to one-quarter of the rates in
England and Wales. Within the USA, the lowest
charges were imposed by the authorities treating
only as far as primary treatment while those
employing secondary treatment were imposing,
depending on the type of effluent, charges up to an
order of magnitude greater.

5.5.2 Monitoring

In terms of the biology of the system, the
degree of treatment required is that which will keep
the load below the assimilative capacity, that is
below the level at which damage will be caused and
the situation should be continually monitored.
Monitoring of the degree of chemical contamination
has already been mentioned (see Chapter Three), but
this alone does not provide an indication of the
biological health of the estuary, and some sort of
assessment procedure along the lines of those in
Chapter Four is necessary.

The kind of monitoring programme required has
usually to be devised to fit the individual
circumstance, and again economics may be an
important factor in the final decision, but the
general pattern would be as depicted in Figure 5.2.

There are basically three ways to decide whether
The degree of contamination has been sufficient to
result in pollution. The first is to make a
comparison with the situation before contamination
started. In most cases this previous data does not

exist, and so the second option is to compare the results with a similar but known unpolluted situation elswhere. Again, because no two locations are identical, this may not be possible, and so the third course of action is to take samples on a gradient from the contamination source and to correlate the degree of biological difference down the gradient with the degree of contamination. Finally even if pollution is demonstrated the need for remedial action may be queried, and Lewis (1978) in a provocative article gives an alternative flowchart for marine pollution studies in which the need for surveys at all is questioned. The reader is also invited to consider whether his suggested end-points (dismiss biologists/shoot industrialist as appropriate) would meet with universal approval!

Figure 5.2 Flowchart of monitoring procedure.

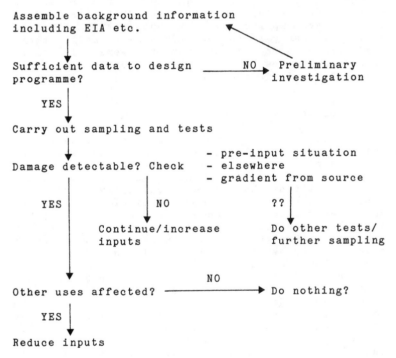

The main problem in such a scheme is to decide on the level of investigation. To take too many samples is costly, time-consuming and wasteful, yet too few may not allow definite conclusions to be drawn. There are four basic ways in which sample sites may be chosen (shown diagrammatically in Figure 5.3):

a) transect - often used in conjunction with a known or suspected contaminant source, e.g. along a shore-line from the mouth of the estuary;
b) grid - in effect a two-dimensional transect, often used for preliminary surveys where nothing is known and complete cover is required;
c) random - preferred statistically as it makes no pre-judgements as to the situation; again often used in new situations;
d) stratified random - where sufficient information exists to be able to divide up the area into different zones (under whatever criteria - pollution status, shore height, sediment type) and within each zone to position the sites at random.

Figure 5.3   Examples of sampling strategy:
              a) transect; b) grid; c) random;
              d) stratified random: the ✷ indicates
              the location of an outfall.

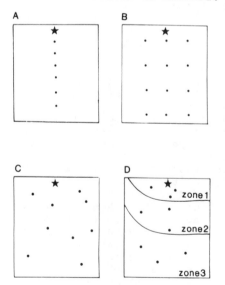

In practice, the stratified random approach has

much to recommend it (for discussion see Hiscock, 1979) as it should ensure that no conditions are unrepresented. This can be important in a typical estuarine situation where the area of pollution may be small. Under the other schemes the result would be a great deal of largely repetitive information from the remainder of the estuary but relatively little from the affected zone itself.

Practical details such as the actual number of samples and the level of investigation depend on the questions being asked and the budget allocated. Wilson and Jeffrey (1987) recommend for their assessment scheme a combination of field biological investigation (the BQI) to establish the amount of damage, with chemical analysis (the PLI) to find the causative agents. Their recommendation was that the investigation should in the preliminary stages at least be kept as simple and as inexpensive as possible, hence their use of shore-based sampling and sediment analysis. Other workers (e.g. Soulsby et al., 1985) have confirmed the advantages, not least in economic terms, of this approach. To this could be added some of the more sensitive sub-lethal assays or specific tests for public health or mariculture if desired.

5.5.3 Fisheries Standards

Fisheries, whether wild or cultured, and amenity are the two other system dependent estuarine uses that have to be considered along with waste assimilative capacity.

The quality required to sustain fisheries is now being specified with increasing stringency. Whereas before it was possible simply to specify an oxygen level (usually 4ppm) sufficient to allow the passage of migratory fish, the environmental quality has increasingly to be such that the flesh of the organisms reared or taken from the estuary meet the standards required for human consumption. These limits may be set by international or other legislation, for example the EEC Shellfish Directive or see Table 5.3 and comments above, or may be established as EQSs through CPA as described in Table 5.1. Milne et al. (1986) give an example of such a procedure for the heavy metal chromium in an estuary where the protection of a shellfishery was of prime importance.

Because of bioaccumulation (in its widest sense), standards for fishery waters may be higher standards may be set for metal contamination.

Saeijs and Stortelder (1982) have given a table of the short- and long-term water quality standards for Lake Grevelingen in the Netherlands for a variety of different functions including conservation, system maintenance, mariculture, fisheries, swimming, agriculture and industry. The actual water quality in the lake was acceptable in most respects for all the different requirements, and were exceeded only with respect to the nutrient concentrations (and also the chloride levels for specific uses such as irrigation). By maintaining these standards all the options are open to the planners and development can proceed in any desired direction.

Fisheries management, in the proper sense of the word, is outside the scope of this volume, but in any case is rarely a problem for the estuarine manager. The exploitation of the stocks is usually regulated by a third party, whether it is the owner of the fishing rights in the case of migratory fish, or some other authority regulating the catch, and the manager's sole responsibility is to meet the required EQO.

The EQSs for Lake Grevelingen in the Netherlands are set out (Table 5.6) in rather more detail than most, both in terms of time-scale and number of parameters.

In the USA for example, the individual States specify only minimum oxygen level (generally 4ppm) and sometimes bacterial contamination limits (mean values around 1,000 - 2,000/100ml). Other contaminants are specified only in terms of a fraction (commonly 0.1 or 0.01 - see e.g. Table 5.4) of the 96h LD_{50} or by reference to levels (unspecified) causing damage or tainting.

Although the two sets of EQSs (Table 5.6) for fresh and salt water are very similar, some concession has ben made to the kind of conditions expected in an estuary, for example no limit specified for suspended sediments in saltwater. The re-suspension of POM from the bottom can add considerably to the BOD in the water column, and this effect will be more marked during spring tides as opposed to neaps. Even without sewage input, such re-suspended natural detritus can add over 2mg BOD/l/day and bring the oxygen saturation down below 60% (Maskell, 1985), clearly raising problems in meeting the relevant EQS. Similarly, re-suspension of bottom deposits can mobilise contaminants bound to the sediment particles, and considerable research is being done at present to try to quantify the rates of these fluxes and their contribution to the overall budget of the system.

Because of the ability of shellish to accumulate pathogens, standards for mariculture waters include additional parameters regarding such contamination. The Lake Grevelingen standards (Saeijs and Stortelder, 1982) for shellfish culture are the same as those for fisheries (Table 5.6) but with the addition of a bacterial standard of 1.5 MPN/l (ST) and 0.5 MPN/l (LT) respectively. The paradox here is of course that the biological system (excluding man) can survive and indeed flourish in areas of high sewage bacterial contamination, athough concern is being expressed at the increasing incidence of botulism in sea gulls.

Table 5.6   Short-term (ST) and long-term (LT) EQSs for fisheries use in Lake Grevelingen, the Netherlands (from Saeijs and Stortelder, 1982)

Parameter	Fresh water		Salt water	
	ST	LT	ST	LT
Oxygen (mg/l)	4-7	5-8	4-7	5-8
Oxygen (% sat)	70-130	100	70-130	100
BOD (mg O2)	5	3	5	3
SS (mg/l)	80	25	-	-
Ammonium (mgN/l)	0.2	0.02	1.0	0.3
Total N (mg/l)	2.5	0.5	1.0	0.3
Total P (mg/l)	0.25	0.05	0.1	0.03
Cd (ug/l)	8	8	8	8
Cr (ug/l)	50	50	50	50
Cu (ug/l)	5	5	5	5
Pb (ug/l)	30	10	30	10
Zn (ug/l)	200	60	200	60
Hg (ug/l)	0.5	0.5	0.5	0.5
Phenol (ug/l)	5	-	5	-

Provided that the pathogens can be either sterilised by cooking or removed by depuration processes, there is no reason why such areas should not be used for mariculture. Not only are they assisting significantly in the assimilation processes of the system, but also their removal results in considerable outputs of what can be regarded essentially as just POM from the system.

This is analogous to the use of beds of reeds or water hyacinths as natural trickling filters to

treat sewage, where the input is converted into a
crop which can then be harvested.
     Within the European Community, member states of
the EEC are required to designate areas for
shellfish culture to meet the standards laid down in
the Shellfish Waters Directive which is displayed in
Table 5.7 with the standards laid down for
Chesapeake Bay by the State of Virginia in the USA.
It is noticeable that the EEC regulations assume
different assimilative capacities for different
waters: where the USA legislation lays down
permitted levels for the water, the EEC standards,
including bacteriological quality, are defined in
terms that take into account bioaccumulation and
bioconcentration by the organisms.

Table 5.7   Selected parameters from the shellfish
             water standards required under Council
             Directive 79/923/EEC (EEC, 1979) and
             Virginia Marine Water Quality Standards
             (EPA, 1978)

Parameter	EEC	Virginia
DO	>80%	4 - 5 mg/l
pH	7-9	6 - 8.5
Total coliforms	-	70/100ml
Faecal coliforms	300 (flesh)	14/100ml
Hg (μg/l)	na	0.1
Cd (μg/l)	na	5.0
Zn	na	0.01 96h LC_{50}
Cu	na	0.1 96h LC_{50}
Hydrocarbons	no surface film	-
PCBs	na	None allowed
DDT	na	1.0 ng/l
Dieldrin	na	3.0 ng/l

na - concentration in water to ensure flesh of
highest marketable quality.

5.5.4 Amenity Standards

     Somewhat surprisingly perhaps, the standards
required for shellfish culture are higher than those
required for amenity usage, even where there is
direct and prolonged contact with the water. Most
States in the USA recognise a heirarchy of degrees
of water contact according to the type of amenity
(see Table 5.2), with the highest quality assigned
to bathing waters. In Europe, designated bathing

beaches have to meet the standards laid down under
Directive 76/160/EEC (Table 5.8).

Table 5.8 Standards required for bathing waters
(EEC, 1976; EPA, 1978); G - guideline,
M - mandatory, ns - not specified:
numbers expressed/100 ml

Parameter	EEC		US EPA
	M	G	
Total coliforms	10,000	500	700
Faecal coliforms	2,000	100	200
Faecal streptococci	300	100	ns
Enteroviruses	0	0	ns
Salmonella	0	0	ns

In addition to the above, standards are also
laid down regarding DO concentration, pH, colour,
odour, transparency, oil films, foams and floating
debris.
Both the EEC and American legislation give
instructions regarding details such as the frequency
of sampling, the kinds of assays to be used and the
absolute limits which much not be exceeded. Most
American States specify not only the mean value
(above) but also the permissible range; for example,
for total coliforms no more than 10% of the samples
can exceed 2,300/100 ml, although here again, their
actual values vary from State to State. In the EEC
legislation, the guideline values are the equivalent
of the EQO, something to be aspired to, as opposed
to the mandatory value which is the upper confidence
limit for the sampling data.

5.6 INTERNATIONAL LEGISLATION

At this point it will have become obvious to the
reader that pollution is becoming increasingly
internationalised, both in terms of its effects and
the action being taken to deal with or to forestall
it. While anyone in the field of estuarine
management will have to work within the framework of
local legislation, this is more and more being
shaped by international agreements, which deserve a
brief mention here.
The prime instruments in this regard are the
Oslo and Paris Conventions - usually referred to
together as the MARPOL Conventions. These owe their
existence, more or less, to an incident involving

the vessel "Stella Maris" which set out from the
Netherlands to dump waste off the coast in April
1971. This attracted so much attention that she was
unable to dump the cargo anywhere, and the publicity
generated by the incident gave such impetus to the
discussions that the Oslo Convention on dumping from
ships at sea was signed and lodged in Oslo in
February of 1972.

The Paris Convention covering discharges from
land based sources followed closely on the heels of
the Oslo Convention, with the result that most
European countries are now signatories to both,
albeit with some delays to enable them to enact the
necessary legislation (Table 5.8).

Table 5.9   Signatories to the Oslo and Paris
             Conventions with dates of implement-
             ation (Oslo and Paris Commissions,
             1984)

Signatory	Oslo	Paris
Norway	1974	1978
Denmark	1974	1978
Sweden	1974	1978
Portugal	1974	1978
Spain	1974	1980
Iceland	1974	1981
France	1974	1978
UK	1975	1978
Netherlands	1975	1978
FRG	1977	1982
Belgium	1978	1984
Finland	1979	-
Ireland	1982	1984
EEC	-	1978

The Oslo and Paris Commissions (1984) offered
(p11) these brief definitions of the two
Conventions:-

"Oslo Convention: the control of a specific source
of potential pollution by a ban on the dumping of
certain substances and the control of all waste
dumping operations".

"Paris Convention: the control of many sources of
pollution by limiting certain substances in
discharges and setting environmental standards, by
monitoring their presence in the marine environment
and by regulating the manufacture or use of the

products in which polluting substances are
contained".

The Oslo Convention stipulates that all dumping
should be licensed, with the added proviso that some
substances should not be dumped at all (Annex I, or
more popularly the "Black List") and others for
which a special prior permit is necessary (Annex II
or "Grey List") (Table 5.10).

Table 5.10 Annex I and Annex II substances under
           the Oslo Convention

Annex I.
1. Organochlorines
2. Mercury and mercury compounds
3. Cadmium and cadmium compounds
4. Persistent plastics and synthetics
5. Hydrocarbons and hydraulic fluids
6. High level radioactivity
7. Materials for biological or chemical warfare
8. Substances which, although capable of being
   degraded, cause tainting or endanger health
9. Does not apply to wastes where the substances
   are present only in trace amounts and which
   therefore come under Annex II or III.

Annex II.
A. Significant amounts of arsenic, lead, copper
   zinc (or their compounds), organosilicons,
   cyanides and fluorides.
B. Beryllium, chromium, nickel and vanadium (and
   their compounds) in acid or alkali waste.
C. Containers, scrap, or bulky waste that might be
   an obstacle to fishing.
D. Radioactive waste not included under Annex I.

Around the same time as the Oslo Convention the
United States was pursuing an initiative on marine
pollution ending with the London Convention on
dumping which incorporated much of the Oslo
agreement, and in 1974 the USA amended the Marine
Protection Research and Sanctuaries Act of 1972 to
implement the London Convention. The United States'
equivalent to the Paris Convention is the Clean
Water Act by which discharges from land-based
sources are regulated and controlled.
     The London Dumping Convention (LDC) itself
entered into force in 1975, and the contracting
countries are required to establish a system to
control and record dumping, to monitor the marine

environment and to make the information available to the Inter-Governmental Maritime Consultative Organisation (IMCO). However, up to the beginning of the 1980's just under half of the contracting countries (mainly the Third World or non-industrialised nations) had not yet provided any information to IMCO, and it is estimated that disposal activities not covered by LDC permits may amount to about 25% of the total.

Just as in the USA the federal legislation is the framework and the umbrella under which State and local policy is enacted, in Europe the EEC, as well as being a signatory to the Paris Convention, was proceeding with a variety of environmental Directives with which the member States would have to comply. Some of these have been mentioned already, but briefly those of most relevance to the estuarine situation are

Directive 76/464/EEC - Dangerous Substances: sets emission standards c.f. the Paris Commission;
Directive 76/160/EEC - Bathing Waters (see Table 5.7);
Directive 78/176/EEC - Titanium Dioxide;
Directive 79/923/EEC - Shellfish Waters (see Table 5.6);
Directive 79/409/EEC - Conservation of Wild Birds;
Directive 82/176/EEC - Hg discharges (chlor-alkali);
Directive 83/513/EEC - Cd discharges
Directive 84/156/EEC - Hg discharges (other than chlor-alkali);

The most comprehensive of these is the Dangerous Substances Directive under which limits are set for discharges from various industries (UES approach), but in addition quality objectives are specified for the aquatic environment (EQO approach). Under this directive the substances banned (List 1 substances) are aldrin, dieldrin, chlordane, hexachloro-cyclohexane (HCH) arsenic, cadmium and mercury with the latter two being covered by subsequent specific directives and four others, namely DDT and its isomers, pentachlorophenol (PCP), carbon tetrachloride and chloroform under discussion for addition to the list.

Other international agreements follow the same pattern. The Helsinki Convention of 1974 for the protection of the Baltic Sea effectively combined the Oslo and Paris Conventions in that it covered all sources of potential pollution, as did the Barcelona Convention of 1976 for the Mediterranean.

Oil pollution has received considerable attention, dating back to the 1954 London Convention (subsequently amended several times) for the prevention of the pollution of the seas by oil. Although the prospect of a catastrophic oil spill is always uppermost in the public mind, the majority of oil entering the sea comes from diffuse and often land-based sources. Operations have been considerably tightened and improved with the introduction of procedures such as the load-on-top (LOT) system to reduce the amount discharged during tank-washing. Nevertheless, the comment would probably be justified that a major part of the oil legislation - and radioactivity would be another example - is concerned more with subjects such as civil liability than with environmental aspects.

The list of EEC directives (above) contains one (conservation of wild birds) which is different in character to the rest in that it specifically imposes an obligation to protect certain species. As with other legislation such as the Ramsar or Bern Conventions for the protection of biota and/or habitats, it can be argued that such an obligation over-rides all others, with the result that discharges or usages which pose a threat but are permitted under other agreements should be curtailed. However, this is rather a contentious point, and in practice it is often difficult to identify exactly what may be posing a danger in such situations.

As well as international agreements, there will be national and local legislation to be taken into account. Because of their situation and character, there are a plethora of different concerns, public and private, with interests to consider. In Dublin Bay, Ireland, there are fifteen authorities including seven Government ministries directly involved, in addition to several private companies and numerous environmental and other action groups. Similarly, in the State of Florida, USA, responsibility and jurisdiction for estuarine management may involve up to ten Federal authorities, along with eleven State, two regional and four local authorities.

In the USA, the major central source for environmental policy is the National Environmental Policy Act of 1969 (as amended). This not only set out the goal of a national policy to protect and

enhance the environment, it provided two principal
safeguards in the requirement for an environmental
impact statement (EIS) for all significant
developments and in the establishment of the right
of private citizens to contest government decisions.
Other important legislation includes programs such
as the Coastal Zone Management Act (1972)
administered through the Department of Commerce,
although the legal background for pollution control
goes back to the Rivers and Harbors Act of 1899.
This has been recently interpreted as a pollution
control measure in that it lays the responsibility
on the Corps of Engineers to maintain the waters not
only so that they are navigable, but also to protect
them for whatever other uses society decides are
desirable. The result is that while the ultimate
responsibility lies with the Department of Defence,
the EPA have a great deal of influence through the
EIS and through its standards with which material
for dumping must comply under the 1972 Act.

In 1987 the Water Quality Act in the USA
formally established the National Estuary Program of
the EPA whose aims are the identification of
nationally significant estuaries, the protection and
improvement of estuarine water quality and the
protection of their living resources. Eleven
estuaries were identified for priority status: four
of these, namely Narragansett Bay in Rhode Island,
Buzzards Bay in Massachussetts, Long Island Sound in
New York and Puget Sound in Washington had already
been studied and assessed under a previous program
in 1985; Albemarle-Pamlico Sounds in North Carolina
and the San Francisco Bay complex in California were
added to the program in 1986 and the five new sites,
Delaware Bay in Delaware and New Jersey, Delaware
Inland Bays in Delaware, the New York/New Jersey
Harbour complex, Sarasota Bay in Florida and
Galveston Bay in Texas.

One of the key elements in the new legislation
is the requirement for management plans to be drawn
up by "management conferences" to include
representatives of the various interests, including
scientists, local, state and federal administrators,
businessmen, farmers, fishermen and the general
public. It will be the duty of the conferences to
set management objectives, such that each area can
decide the optimum strategy for their particular
situation, and to monitor the progress made toward
these objectives.

Mackay (1985) in his comments on estuarine
management drew attention to one of the main

problems of management in that there may be a
conflict of interest in the administration. In
England and Wales, control is in the hands of the
Water Authorities, who also take responsibility for
sewage disposal although consent for the discharges
has to be obtained from the Secretary of State, and
this situation was contrasted with that in Scotland
where control, through the River Purification
Boards, is independent of the drainage authorities.

Mackay (1985) also highlighted the negation of
environmental legislation by the authorities. Taking
as the example the EEC Bathing Waters and Shellfish
Waters Directives, he showed how financial criteria
had been set for their application with the result
that only waters which already met the standards
were designated, irrespective of their actual or
potential usage (no Scottish beaches were designated
under the Bathing Waters Directive yet see for
example Figure 1.2).

Other countries can quote similar examples.
There is no need for an EIS if the development is
not regarded by the U.S. Corps of Engineers as
significant, and in these cases, public
participation in the planning process either through
litigation or planning hearings has often been a
crucial element in the process.

There is a very fine distinction to be made
here, as between tax avoidance, which is legal and
even held to be praiseworthy, and tax evasion, which
is neither. It is perfectly possible for authorities
to observe the letter of the law without observing
the spirit, and environmental legislation is in
general open to considerable legal ambiguity.
Considering the MARPOL Conventions above, the
definition of "significant amounts", "high level"
(radioactivity) and "trace amounts" depends
largely on the individual viewpoint. Does dilution
in itself suffice to change a waste from Annex I
(significant, high level) to Annex II (trace, low
level) or should the total quantity discharged be
taken into account? The latter question would be
best answered by the EQO approach, whereby it is the
ability of the system to deal with a contaminant
that determines how much can be discharged and not
just its concentration, but then how to consider the
situation of nuclear reprocessing, where small
amounts of high level waste are taken in and large
amounts of low level waste are discharged?

5.7 FUTURE DEVELOPMENTS

The nuclear reprocessing situation encapsulates many of the environmental issues. The effects can be detected far beyond the boundaries of the host country, there are wide differences of opinion over what may be considered "safe" levels, and political considerations through the national energy or defence policy must be considered. It also provides a good example of a growing trend of exporting pollution, in that a nuclear waste reprocessing plant may receive waste from other countries unwilling or unable to deal with it themselves. In this context mention has already been made of the exploitation of the Third World and the lack of environmental legislation there.

In the industrialised world there is a growing realisation that the amounts of waste discharged into the environment must be reduced and that conservation and environmental protection do have a place in society. However, at the individual level, it is difficult for people to accept what they see as a decline in their own standard of living, in that to accommodate an environmental perspective they personally will be losing more than they will be gaining. Likewise it is difficult to urge the under-developed nations to curtail or cancel developments which they see as necessary to bring their living standards up to those they see in other countries.

The economic trend at the moment in the developed nations is toward less and less heavy industry, and this combined with the decline and change in shipping in many estuaries has led to a big change in the loads imposed on many estuaries. In addition, population growth, and hence city growth, has halted in most countries in North America and Europe (with one or two exceptions).

If one compares the example of the Clyde estuary today with the situation at the turn of the century, or even thirty years ago, steelmaking and shipbuilding are operating at a fraction of what they were, and the population of the city of Glasgow is barely two-thirds of what it was in its heyday. Indeed it could be argued that improvement in many cases owes as much to these changes as to any actions by management.

Likewise, traditional fishing has declined and in its place mariculture is assuming an increasingly important role. Mariculture demands a clean environment, yet there are signs that it in turn is threatening its own survival. Some of the concerns that have been voiced include eutrophication and de-oxygenation below fish-rearing cages and the effects of some of the compounds and preparations used (including until recently TBT as anti-fouling on the nets) on both the environment and the consumer.

Sport fishing is gaining in importance in many local economies, and amenity usage of estuaries in general is increasing. In many cities, disused docks, whose water quality while perhaps improving would be unlikely to gain general acceptance for either contact watersports or fishing, are being converted into marinas to cater for pleasure boating.

Needless to say, even amenity usages can conflict. Birds are particularly sensitive to disturbance and on certain beaches in the U.K., zones have been declared out of bounds for bait-digging to protect the birds, and successful prosecutions brought against people infringing the ban. Recent research has also highlighted situations in which the suggestion is that a degree of pollution may be necessary for the maintenance of the bird populations (Merne, 1985; van Impe, 1985).

In many cases it is the perception of the pollution status as much as the actual status which will determine what is done. Thus it is important firstly to involve as many interested parties as possible, as the USA example is doing, and to ensure that any course of action is kept as flexible as possible, such that future changes in priorities and objectives can be accommodated.

A recent survey throughout Europe has found that fewer than one in ten people ranked economic development above environmental protection (although fewer than one in five thought that their government was doing the correct things about the environmental situation)! The major issues on a global scale were the safe disposal of industrial waste and prevention of damage to sea life and beaches (coming before issues such as air pollution, animal extinction and the loss of natural resources) (EEC, 1986).

On other continents the situation may be different. In many countries populations and cities are increasing in size, and will continue to do so for the foreseeable future. Coupled with this are signs of deforestation and erosion, the effects of which are transferred down to estuaries, leading in many places to the smothering and breakdown of coral reefs. On top of this is the trend toward growth of heavy industry, with the result that many places

find themselves with similar problems to European
cities at the turn of the century but with far less
resources to deal with the problem. It is perhaps to
here that future efforts should be directed.

BIBLIOGRAPHY

Abram, J.W. and Nedwell, D.B. (1978) 'Inhibition of
    methano-genesis by sulphate reducing bacteria
    competing for transferred hydrogen'. Arch.
    Microbiol. 117, 89 - 92.
Ademoroti, C.M.A. (1984) 'Short term BOD tests'.
    Effl. Wat. Treat. J. 24(10), 373 - 377.
Ademoroti, C.M.A. (1986) 'Models to predict BOD from
    COD values'. Effl. Wat. Treat. J. 26(3), 80 -
    84.
Allen, J.R.L. and Rae, J.E. (1986) 'Time sequence of
    metal pollution in the Severn estuary, southwest
    UK'. Mar. Pollut. Bull. 17, 427 - 431.
Anderson, J.G. and Meadows, P.S. (1978)
    'Microenvironments in marine sediments'. Proc.
    R. Soc. Edinb. 76B, 1 - 16.
Andrews, M.J. (1984) 'Thames estuary: pollution and
    recovery'. in P.J. Sheehan, D.R. Miller, G.C.
    Butler and P. Bourdeau (eds.) 'Effects of
    Pollution at the Ecosystem Level', SCOPE 22,
    John Wiley and Sons, Chichester, pp. 195-227.
Andrews, M.J. and Richard, D. (1980) 'Rehabilitation
    of the inner Thames estuary'. Mar. Pollut. Bull.
    11, 327 - 331.
Ansell, A.D., Barnett, P.R.O., Bodoy, A. and Masse,
    H. (1980) 'Upper temperature tolerances of some
    European molluscs'. Mar. Biol. 58, 33 - 39.
Arenas, V.F. and de la Lanza, G.E. (1983) 'Annual
    phosphorus budget of a coastal lagoon in the
    northwest of Mexico' in R. Hallberg (ed.)
    'Environmental Biogeochemistry', Ecol. Bull.
    (Stockholm) 35, 431 - 440.
Aston, S.R. (1986) 'Development, testing and inter-
    calibration of reference methods for pollution
    studies in coastal and estuarine waters'. Wat.
    Sci. Technol. 18, 27 - 34.
Atkinson, L.P. and Hall, J.R. (1976) 'Methane dis-

tribution and production in the Georgia salt marsh'. Est. Coast. Mar. Sci. 4, 677 - 686.

Baird, D. and Milne, H. (1981) 'Energy flow in the Ythan estuary, Aberdeenshire, Scotland'. Estuar. Coast. Mar. Sci. 13, 455 - 472.

Barnes, R.S.K. (1979) 'Intrapopulation variation in *Hydrobia* sediment preferences'. Estuar. Coast. Mar. Sci. 9, 231 - 234.

Bayne, B.L. (1985) 'Responses to environmental stress: tolerance, resistance and adaptation' in J.S. Gray and M.E. Christiansen (eds.) 'Marine Biology of Polar Regions and Effects of Stress on Organisms', Proceedings of the 18th EMBS, Oslo 1983, John Wiley and Sons, Chichester, pp. 331 - 349.

Bayne, B.L., Brown, D.A., Burns, K., Dixon, D.R., Ivanovici, A., Livingstone, D.R., Lowe, D.M., Moore, M.N., Stebbing, A.R.D. and Widdows, J. (1985) 'The Effects of Stress and Pollution on Marine Animals'. Praeger Scientific, New York.

Bayne, B.L., Moore, M.N., Widdows, J., Livingstone, D.R. and Salkeld, P. (1979) 'Measurement of the responses of individuals to environmental stress and pollution: studies with bivalve molluscs' Phil. Trans. R. Soc. Lond. B, 286, 563 - 581.

Bayne, B.L. and Scullard, C. (1977) 'Rates of nitrogen excretion by species of *Mytilus* (Bivalvia: Mollusca). J. mar. biol. Ass. U.K., 57, 355 - 369.

Bellamy, J.D., John, D.M. and Whittick, A. (1968) 'The "kelp forest ecosystem" as a "phytometer" in the study of pollution in the inshore environment' Rep. Underwat. Ass. 79 - 82.

Bengtsson, B.-E. 'Biological variables, especially skeletal deformities in fish for monitoring marine pollution'. Phil. Trans. R. Soc. Lond. B 286, 457 - 464.

Biggs, R.B. (1970) 'Sources and distribution of suspended sediment in North Chesapeake Bay'. Mar. Geol. 9, 187 - 201.

Birbeck, T.H. and McHenery, J.G. (1982) 'Degradation of bacteria by *Mytilus edulis*'. Mar. Biol. 72, 7 - 15.

Blackstock, J. (1980) 'A biochemical approach to assessment of the effects of organic pollution on the metabolism of the non-opportunistic polychaete, *Glycera alba*'. Helgolander Meeresunters., 33, 546 - 555.

Blackstock, J. (1984) 'Biochemical metabolic regulatory responses of marine invertebrates to natural environmental change and marine pollution'. Oceanogr. Mar. Biol. Ann. Rev. 22, 263 - 314.

Bliss, C.I. (1935) 'The calculation of the dosage mortality curve'. Ann. Appl. Biol. 22, 134 - 167.

Boaden, P.J.S. (1977) 'Thiobiotic facts and fancies (aspects of the distribution and evolution of anaerobic meiofauna)'. Mikrofauna Meeresbod. 61, 45 - 63.

Boelens, R.G. (1987) 'The use of fish in water pollution studies' in D.H.S. Richardson (ed.) 'Biological Indicators of Pollution', Royal Irish Academy, Dublin, pp. 89 - 110.

Bryan, G.W. and Hummerstone, L.G. (1971) 'Adaptation of the polychaete *Nereis diversicolor* to estuarine sediments containing high concentrations of heavy metals. 1. General observations and adaptation to copper'. J. mar. biol. Ass. U.K. 51, 845 - 863.

Bryan, G.W., Langston, W.J. and Hummerstone, L.G. (1980) 'The use of biological indicators of heavy metal contamination in estuaries'. Mar. biol. Ass. U.K. Occ. Publ. No. 1, 1 - 73.

Bryan, G.W., Langstone, W.J., Hummerstone, L.G. and Burt, G.R. (1985) 'A guide to the assessment of heavy-metal contamination in estuaries using biological indicators'. Mar. biol. Ass. U.K. Occ. Publ. No. 4, 1 - 92.

Burke, M.V. and Mann, K.H. (1974) 'Productivity and production: biomass ratios of bivalve and gastropod populations in an eastern Canadian estuary'. J. Fish. Res. Bd. Can. 31, 167 - 177.

Calmano, W., Wellershaus, S. and Liebsch, H. (1984). 'The Weser estuary: a study on heavy metal behaviour under hydrographic and water quality conditions'. Veroff. Inst. Meeresforsch. Bremerh. 20, 151 - 182.

Cameron, W.M. and Pritchard, D.W. (1963) 'Estuaries' in M.N. Hill (ed.) 'The Sea' vol. 2, John Wiley and Sons, New York, pp. 306 - 324.

Cammen, L.M. (1976) 'Accumulation time and turnover rate of organic carbon in salt marsh sediment'. Limnol. Oceanogr. 20, 1012 - 1015.

Capuzzo, J.M. (1985) 'The effects of pollutants on marine zooplankton populations: sublethal physiological indices of pollution stress' in J.S. Gray and M.E. Christiansen (eds.) 'Marine Biology of Polar Regions and Effects of Stress on Marine Organisms', Proceedings of the 18th EMBS, Oslo 1983, John Wiley and Sons, Chichester, pp 475 - 491.

Carricker, M.R. (1967) 'Ecology of estuarine benthic invertebrates' in G.H. Lauff (ed.) 'Estuaries', Publ. No. 83, A.A.A.S., Washington, D.C., pp. 442 - 487.

Carson, R. (1962) 'Silent Spring'. Houghton Mifflin, Boston.

Chambers, M.R. and Milne, H. (1975a) 'Life cycle and production of *Nereis diversicolor* in the Ythan estuary, Scotland'. Estuar. Coast. Mar. Sci. 3, 133 - 144.

Chambers, M.R. and Milne, H. (1975b) 'The production of *Macoma balthica* in the Ythan estuary'. Estuar. Coast. Mar. Sci. 3, 443 - 455.

Chapman, A.G., Fall, L. and Atkinson, D.E. (1971) 'Adenylate energy charge in *Escheriscia coli* during growth and starvation'. J. Bact. 108, 1072 - 1086.

Chapman, P.M., Dexter, R.N., Cross, S.F. and Mitchell, D.G. (1986) 'A field trial of the sediment quality triad in San Francisco Bay'. NOAA Technical Memorandum NOS OMA 25, 1 - 134.

Clark, R.B. (1987) 'The Waters around the British Isles'. Clarendon Press, Oxford.

Connell, D.W. (1982) 'An approximate hydrocarbon budget for the Hudson Raritan estuary - New York'. Mar. Pollut. Bull. 13, 89 - 93.

Cruz, de la, A.A. and Gabriel, R.C. (1974) 'Caloric, elemental and nutritive changes in decomposing *Juncus* leaves'. Ecology 55, 882 - 886.

Curds, C.R. and Hawkes, H.A. (1983) 'Ecological Aspects of Wastewater Treatment'. 2 vols. Academic Press, London.

Curran, J.C. and Milne, D.P. (1985) 'Development of a model of bacterial pollution in the Clyde estuary' in J.G. Wilson and W. Halcrow (eds.) 'Estuarine Management and Quality Assessment', Plenum Press, London, 37 - 50.

Dame, R.F. (1976) 'Energy flow in an intertidal oyster population'. Estuar. Coast. Mar. Sci. 4, 243 - 253.

Davenport, J. (1985) 'Environmental Stress and Behavioural Adaptation' Croom Helm, London.

Davies, J.M. and Gamble, J.C. (1979) 'Experiments with large enclosed ecosystems'. Phil. Trans. R. Soc. Lond. B 286, 523 - 544.

Davis, J.P. and Wilson, J.G. (1985) 'The energy budget and population structure of *Nucula turgida* in Dublin Bay'. J. Anim. Ecol. 54, 557 - 571.

Deering, N. and Gray, N.F. (1987) 'The polluter pays principle: a comparison of charging systems in

Europe and the USA'. Water Technology Research, University of Dublin, TR 3.

Delaune, R.D., Hambrick, G.A. and Patrick, W.H. (1980) 'Degradation of hydrocarbons in oxidised and reduced sediments'. Mar. Pollut. Bull. 11, 103 - 106.

Dethlefsen, V. (1984) 'Diseases in North Sea fishes'. Helgolander Meeresunters. 37, 353 - 374.

Dixon, D.R. (1983) 'Sister chromatid exchange and mutagens in the marine environment'. Mar. Pollut. Bull. 14, 282 - 284.

Dixon, D.R. and Clarke, K.R. (1982). 'Sister chromatid exchange, a sensitive method for detecting damage caused by environmental mutagens in the chromosomes of adult *Mytilus edulis*'. Mar. Biol. Letters 3, 163 - 172.

Dixon, D.R. and Pollard, D. (1985) 'Embryo abnormalities in the periwinkle, *Littorina "saxatilis"*, as indicators of stress in polluted marine environments. 'Mar. Pollut. Bull. 16, 29 - 33.

Drew, E.A. (1971) 'Botany' in J.D. Woods and J.L. Lythgoe 'Underwater Science', Oxford University Press, London.

EEC (1976) 'A Council directive on the quality required of bathing waters'. Council Directive 76/160/EEC, Brussels.

EEC (1979) 'A Council Directive on the quality required of shellfish waters'. Council Directive 79/923/EEC, Brussels.

EEC (1986) 'The European and their environment in 1986'. EEC, Brussels.

Elliott, M. and Griffiths, A.H. (1986) 'Mercury contamination in components of an estuarine ecosystem'. Wat. Sci. Technol. 18, 161 - 170.

Elliott, M. and McLusky, D.S. (1985) 'Invertebrate production ecology in relation to estuarine management' in J.G. Wilson and W. Halcrow (eds.) 'Estuarine Management and Quality Assessment', Plenum Press, London, pp. 85 - 104.

EPA (1978) 'A compilation of state water quality standards for marine waters'. U.S. Environmental Protection Agency, Washington, D.C.

EPA (1987) 'Protecting our estuaries'. EPA Journal 13.6, July/ August 1987, U.S. Environmental Protection Agency, Washington, D.C.

Fabris, J.G., Smith, K.A., Atack, J.E., Hefter, G. and Kilpatrick, A.L. (1986) 'Submersible integrating water sampler for heavy metals', Wat. Res. 20, 1393 - 1396.

Fenchel, T.M. and Blackburn, T.H. (1979) 'Bacteria and Mineral Cycling'. Academic Press, London.

Fenchel, T.M. and Reidl, R.J. (1970) 'The sulphide system: a new biotic community underneath the oxidised layer of marine sand bottoms'. Mar. Biol. 7, 255 - 268.

Gameson, A.L.H. (1985) 'Application of coastal pollution research. Part 5. Microbial mortality'. Water Research Centre, Medmenham, TR 228.

Gardiner, J. and Mance, G. (1984) 'Environmental quality standards for List 11 substances'. Water Research Centre, Medmenham, TR 206.

Garton, D.W. and Stickle, R.B. (1985) 'Relationship between multiple locus heterozygosity and fitness in the gastropods *Thais haemastoma* and *Thais lamellosa* ' in J.S. Gray and M.E. Christiansen (eds.) 'Marine Biology of Polar regions and Effects of Stress on Marine Organisms', Proceedings of the 18th EMBS, Oslo 1983, John Wiley and Sons, Chichester, pp. 544 -554.

Germano, J.D. (1983) 'High resolution sediment profiling with REMOTS camera system'. Sea Technol., December, 1983.

GESAMP (1980) 'Monitoring biological variables related to marine pollution'. Rep. Stud. GESAMP (12): 22pp.

GESAMP (1982) 'Scientific criteria for the selection of waste disposal sites at sea'. Rep. Stud. GESAMP (16): 60pp.

GESAMP (1984a) 'Marine pollution implications of ocean energy development'. Rep. Stud. GESAMP (20): 44pp.

GESAMP (1984b) 'Review of potentially harmful substances - cadmium, lead and tin'. Rep. Stud. GESAMP (22): 114pp.

GESAMP (1986) 'Environmental capacity: an approach to marine pollution prevention'. Rep. Stud. GESAMP (30): 49pp.

Gibbs, P.E. and Bryan, G.W. (1986) 'Reproductive failure in populations of the dog-whelk, *Nucella lapillus*, caused by imposex induced by tributyltin from antifouling paints'. J. mar. biol. Ass. U.K. 66, 767 - 777.

Godshalk, G. and Wetzel, R.G. (1978) 'Decomposition of aquatic angiosperms. 111. *Zostera marina* L. and a conceptual model of decomposition'. Aquat. Bot. 5, 329-354.

Goldberg, E.W., Bowen, U.T., Farrington, J.W., Harvey, G., Martin, J.H., Parker, P.L.,

Riseborough, R.W., Robertson, W., Schneider, E. and Gamble, E. (1978) 'The mussel watch'. Environ. Conserv. 5, 1 - 25.

Grassle, J.F. and Grassle, J.P. (1974) 'Opportunistic life histories and genetic systems in marine benthic polychaetes'. J. Mar. Res. 32, 253 - 284.

Gray, J.S. (1976) 'The fauna of the polluted River Tees estuary'. Estuar. Coast. Mar. Sci. 4, 653 - 676.

Gray, J.S. (1979) 'Pollution-induced changes in populations'. Phil. Trans. R. Soc. Lond. B 286, 545 - 561.

Gray, J.S. (1981) 'The Ecology of Marine Sediments', Cambridge University Press, Cambridge.

Gray, J.S. and Mirza, F.B. (1979) 'A possible method for detecting pollution-induced disturbance on marine benthic communities'. Mar. Pollut. Bull. 10, 142 - 146.

Gray, J.S. and Pearson, T.H. (1982) 'Objective selection of sensitive species indicative of pollution-induced change in benthic communities. 1. Comparative methodology'. Mar. Ecol. Prog. Ser. 9, 111 - 119.

Gray, N.F. (1986) 'The bathing water directive - a challenge for Ireland'. Tech. Ire. December 1986, 15 - 19.

Gray, N.F. (1988) 'The Biology of Wastewater Treatment'. Oxford University Press, Oxford.

Gulliksen, B., Haug, T. and Sandnes, O.K. (1980) 'Benthic macrofauna on new and old grounds at Jan Mayen'. Sarsia 65, 137 - 148.

Hartwig, E.O. (1976) 'The impact of nitrogen and phosphorus release from a siliceous sediment on the overlying water' in M. Wiley (ed.) 'Estuarine Processes', Academic Press, New York, pp. 103 - 117.

Harvey, G.R., Miclas, H.P., Bowen, V.T. and Steinhauser, W.G. (1974) 'Observations on the distribution of chlorinated hydrocarbons in Atlantic Ocean organisms'. J. Mar. Res. 32, 103 - 118.

Heald, E.J. (1969) 'The production of organic detritus in a south Florida estuary'. Ph.D. Thesis, University of Miami, Florida.

Hedgepeth, J.W. (1983) 'Brackish waters, estuaries and lagoons' in O. Kinne (ed.) 'Marine Ecology' Vol. 5 Part 2, John Wiley and Sons, Chichester, 739 - 793.

Heidinger, R.C. and Crawford, S.D. (1977) 'Effects of temperature and feeding rate on the liver/somatic index of the largemouth bass *Micropterus salmoides*'. J. Fish. Res. Bd. Can. 34, 633 - 638.

Heip, C. and Herman, R. (1979) 'Production of *Nereis diversicolor* O.F. Muller (Polychaeta) in a shallow brackish-water pond'. Estuar. Coast. Mar. Sci. 8, 297 - 305.

Henriksen, K. and Jensen, A. (1979) 'Nitrogen mineralisation in a salt marsh ecosystem dominated by *Halimione portulacoides*' in R.L. Jefferies and A.J. Davy (eds.) 'Ecological Processes in Coastal Environments', Blackwell, Oxford, pp. 373 - 384.

Hibbert, C.J. (1976) 'Biomass and production of a bivalve community on intertidal mudflats' J. exp. mar. Biol. Ecol. 25, 249 - 261.

Hiscock, K. (1979) 'Systematic surveys and monitoring in nearshore sub-littoral areas using diving' in D. Nichols (ed.) 'Monitoring the Marine Environment', Institute of Biology, London, pp. 55 - 74.

Hughes, R.N. (1970) 'An energy budget for a tidal flat population of the bivalve *Scrobicularia plana* (da Costa)'. J. Anim. Ecol. 39, 357 - 381.

Hunt, G.J. (1987) 'Radioactivity in surface and coastal waters of the British Isles, 1986'. Aquat. Environ. Monit. Rep., MAFF Direct. Fish. Res., Lowestoft, 18, 1 - 61.

ICRP (1977) 'Recommendations of the International Commission on Radiological Protection'. ICRP Publication No. 26, Pergamon Press, Oxford.

Ivanovici, A.M. (1980) 'Application of adenylate energy charge to problems of environmental impact assessment'. Helgolander Meeresunters. 33, 556 - 565.

Jefferies, R.L. (1972) 'Aspects of salt marsh ecology with particular reference to inorganic plant nutrition' in R.S.K. Barnes and A.J. Green (eds.) 'The Estuarine Environment', Applied Science Publishers, London, pp. 61 - 85.

Jeffrey, D.W., Harris, C.R., Tomlinson, D.L. and Wilson, J.G. (1980) 'A manual for the assessment of estuarine quality'. National board of Science and Technology, Dublin.

Jeffrey, D.W., Wilson, J.G., Harris, C.R. and Tomlinson, D.L. (1985) 'The application of two simple indices to Irish estuary pollution status' in J.G. Wilson and W. Halcrow (eds.) 'Estuarine Management and Quality Assessment', Plenum Press, London, pp. 147 - 162.

Jeffries, H.P. (1972) 'A stress syndrome in the hard

clam, *Mercenaria mercenaria*'. J. Invert. Pathol. 20, 242 - 251.

Jensen, K. (1983) 'Sources and routes of removal of petroleum hydrocarbons in a Danish marine inlet'. Oil. Petrochem. Pollut. 1, 207 - 216.

Johnston, R. (1983) 'Environmental quality assessment in estuaries and coastal waters'. ICES Coop. Res. Rept. 118, 109 - 118.

Johnston, R. (1984) 'Oil pollution and its management' in O. Kinne (ed.) 'Marine Ecology' Vol. 5, John Wiley and Sons, Chichester, pp. 1433 - 1582.

Joint, I.R. (1978) 'Microbial production of an estuarine mudflat'. Estuar. Coast. Mar. Sci. 7, 185 - 195.

Jones, J.A. (1968) 'Primary productivity of the tropical marine turtlegrass *Thalassia testudinum* Konig and its epiphytes'. Ph.D. Thesis, University of Miami, Florida.

Jones, K. (1974) 'Nitrogen fixation in a salt water marsh'. J. Ecol. 62, 553 - 565.

Jones, N.A. and Wolff, W.J. (eds.) (1981) 'Feeding and Survival Strategies of Estuarine Organisms'. Plenum Press, New York.

Kamlet, K.S. (1983) 'Dredged-material ocean dumping: perspectives on legal and environmental impacts' in D.R. Kester, B.H. Ketchum, I.W. Duedall and P.K.Park (eds.) 'Wastes in the Ocean', Volume 2, John Wiley and Sons, New York, 30 - 65.

Kanwisher, J.W. (1966) 'Photosynthesis and respiration in some seaweeds' in H. Barnes (ed.) 'Some Contemporary Studies in Marine Science', Allen and Unwin, London, pp. 407 - 420.

Kanwisher, J.W. and Wainwright, S.A. (1967) 'Oxygen balance in some reef corals'. Biol. Bull. marine Biol. Lab., Woods Hole, 133, 378 - 90.

Kerstner, M. and Forstner, U. (1986) 'Chemical fractionation of heavy metals in anoxic estuarine and coastal sediments'. Wat. Sci. Technol. 18, 121 - 130.

Kester, D.R., Ketchum, B.H., Duedall, I.W. and Park, P.K. (eds.) (1983) 'Dredged-material disposal in the ocean'. 'Wastes in the Ocean', Volume 2, John Wiley and Sons, New York.

Kirkman, H. and Reid, D.D. (1979) 'A study of the role of a seagrass *Posidonia australis* in the carbon budget of an estuary'. Aquat. Bot. 7(2), 173 - 183.

Klinowska, M. (1986) 'The status of marine mammals in the North Sea' in 'Reasons for Concern', Proceedings of the 2nd North Sea Seminar,

Rotterdam 1986, Werkgroep Nordzee, Amsterdam.

Krebs, C.J. (1978) 'Ecology: the Experimental Analysis of Distribution and Abundance', 2nd edition, Harper International, New York.

Kuenzler, E.J. (1961) 'Structure and energy flow of a mussel population in a Georgia salt-marsh'. Limnol. Oceanogr. 6, 191 - 204.

Laws, E.A. (1981) 'Aquatic Pollution', Wiley Interscience, London.

Lee, R.F., Stolzenbach, J., Singer, S. and Tenore, K.R. (1981) 'Effects of crude oil on growth and mixed function oxygenase activity in polychaetes, *Nereis* sp.' in J. Vernberg, J. Calabrese, F.P. Thurberg and W.B. Vernberg (eds.) 'Biological Monitoring of Marine Pollutants', Academic Press, New York, pp. 323 - 334.

Leentvaar, J and Nijboer, S.M. (1986) 'Ecological impacts of the construction of dams in an estuary'. Wat. Sci. Tech. 18, 181 - 191.

Leppakowski, E. (1977) 'Monitoring the benthic environment of organically polluted river mouths' in J.S. Alabaster (ed.) 'Biological Monitoring of Inland Fisheries', Applied Science Publishers, Barking, pp. 125 - 132.

Lewis, J.R. (1978) 'The implications of community structure for benthic monitoring studies'. Mar. Pollut. Bull. 9, 64 - 67.

Lichtfuss, R. and Brummer, G. (1977) 'Schwertmetall-belastung von Elbe-sedimenten'. Naturwissen-schaften 64, 122 - 125.

Long, E.R. and Chapman, P.M. (1985) 'A sediment quality triad: measures of sediment contamination, toxicity and infaunal community composition in Puget Sound'. Mar. Pollut. Bull. 16, 405 - 416.

Loring, D.H. and Prosi, F. (1986) 'Cadmium and lead cycling between water, sediment and biota in an artificially contaminated mud flat on Borkum (FRG)'. Wat. Sci. Technol. 18, 131 - 140.

Lowe, D.M., Moore, M.N. and Clarke, K.R. (1981). 'Effects of oil on digestive cells in mussels: quantitative alterations in cellular and lysosomal structure'. Aquat. Toxicol. 1, 213 - 226.

Lowe, D.M. and Pipe, R.K. (1985) 'Cellular responses in the mussel *Mytilus edulis* following exposure to diesel oil emulsions: reproductive and nutrient storage cells'. Mar. Environ. Res. 17, 234 - 237.

Lowe, D.M. and Pipe, R.K. (1987) 'Mortality and

quantitative aspects of storage cell utilisation in mussels *Mytilus edulis* following exposure to diesel oil hydrocarbons'. Mar. Environ. Res. 22, 243 - 52.

Luoma, S.N. and Bryan, G.W. (1982) 'A statistical study of environmental factors controlling concentrations of heavy metals in the burrowing bivalve *Scrobicularia plana* and the polychaete *Nereis diversicolor*'. Estuar. Coast. Shelf Sci. 15, 95 - 108.

Mackay, D.W. (1985) 'Progress in estuarine water quality management: an overview' in J.G. Wilson and W. Halcrow (eds.) 'Estuarine Management and Quality Assessment', Plenum Press, London, pp. 163 - 172.

Mackay, D.W., Haig, A.J.N. and Allcock, R. (1986) 'Licensing a major industrial discharge to coastal waters. The practical application of the EQO/EQS approach'. Wat. Sci. Technol, 18 287 - 295.

Mackay, D.W., Taylor, W.K. and Henderson, A.R. (1978) 'The recovery of the polluted Clyde estuary'. Proc. R. Soc. Edinb. B 76, 135 - 152.

McElroy, A.E. (1985) 'Physiological and biochemical effects of the polycyclic aromatic hydrocarbon benz(α)anthracene on the deposit feeding polychaete *Nereis virens*' in J.S. Gray and M.E. Christiansen (eds.) 'Marine Biology of Polar Regions and Effects of Stress on Marine Organisms', Proceedings of the 18th EMBS, Oslo 1983, John Wiley and Sons, Chichester, pp. 527 - 543.

McLusky, D.S. (1981) 'The Estuarine Ecosystem'. Blackie, Glasgow.

McLusky, D.S., Bryant, V. and Campbell, R. (1986) 'The effects of temperature and salinity on the toxicity of heavy metals to marine and estuarine invertebrates'. Oceanogr. Mar. Biol. Ann. Rev. 24, 481 - 520.

Madden, B. (1984) 'The Nitrogen and Phosphorus Turnover of the *Salicornia* flat at North Bull Island, Dublin Bay'. B.A. Mod. Thesis, Trinity College Dublin.

Magennis, B. (1987a) 'Hydroids and estuarine water quality' in D.H.S. Richardson (ed.) 'Biological indicators of pollution', Royal Irish Academy, Dublin, pp. 211 - 224.

Magennis, B. (1987b) 'The Assessment of Pollution at the Bull Island, Dublin'. Ph.D. Thesis, University of Dublin.

Mance, G. (1987) 'Pollution Threat of Heavy Metals

in the Aquatic Environment'. Elsevier Applied Science, London.

Mann, K.H. (1982) 'Ecology of Coastal Waters', Blackwell, Oxford.

Martin, J.M., Mouchel, J.M. and Nirel, P. (1986) 'Some recent developments in the characterisation of estuarine particulates'. Wat. Sci. Technol. 18, 83 - 92.

Martin, R.G. (1977) 'PCBs - polychlorinated biphenyls'. Sport Fish. Inst. Bull. No. 288, 1 - 3.

Maskell, J.M. (1985) 'The effect of particulate BOD on the oxygen balance of a muddy estuary' in J.G. Wilson and W. Halcrow (eds.) 'Estuarine Management and Quality Assessment', Plenum Press, london, pp. 51 - 60.

Meadows, P.S. and Campbell, J.I. (1978) 'An Introduction to Marine Science', Blackie, Glasgow.

Mechelas, B.J. (1974) 'Pathways and environmental requirements for biogenic gas production in the oceans'. In I.R. Kaplan (ed) 'Natural Gases in Marine Sediments', Plenum Press, New York, pp. 12 - 25.

Meredith, W.H. and Lotrich, V.A. (1979) 'Production dynamics of a tidal creek population of *Fundulus heteroclitus* (L.)'. Estuar. Coast. Mar. Sci. 8, 99 - 118.

Merne, O. (1985) 'Macrofauna and their availability as a food resource for birds in the Shannon and Fergus estuaries'. M.Sc. Thesis, University of Dublin.

Mertens, E.W. and Gould, J.R. (1979) 'The effects of oil on marine life'. Erdol & Kohle Erdgas Petrochem. 32, 162 - 166.

Milne, H. and Dunnet, G. (1972) 'Standing crop, productivity and trophic relations of the fauna of the Ythan estuary' in R.S.K. Barnes and J. Green (eds.) 'The Estuarine Environment', Applied Science Publishers, London, pp. 86 - 106.

Milne, R.A., Nicholas, P.C., Pattinson, C. and Halcrow, W. (1986) 'The definition of effluent discharge consent conditions in complex estuarine environments'. Wat. Sci. Technol. 18, 267 - 276.

Mitchell, P.I., Sanchez-Cabeza, J.A., Clifford, H., Vidal-Quadras, A. and Font, J.L. (1986) 'The distribution of radiocaesium, radioiodine and plutonium around the Irish coastline using *Fucus vesiculosus* as a bio-indicator' in D.H.S.

Richardson (ed.) 'Biological Indicators of Pollution', Royal Irish Academy, Dublin, pp. 1 - 12.

Mitchell, R. (1974) 'Introduction to Environmental Microbiology', Prentice Hall, New Jersey.

Moore, M.N., Lowe, D.M., Livingstone, D.R. and Dixon, D.R. (1986) 'Molecular and cellular indices of pollution and their use in environmental impact assessment'. Wat. Sci. Technol. 18, 223 - 232.

Morel, F.M.M. and Schiff, S.L. (1983) 'Geochemistry of municipal waste in coastal waters' in E.P. Myers (ed.) 'Ocean Disposal of Municipal Waste Waters: impacts on the coastal environment', Massachussetts Institute of Technology, Cambridge, Mass., pp. 251 - 421.

Morris, A.W., Bale, A.J. and Howard, R.J.M. (1982) 'The dynamics of estuarine manganese cycling'. Estuar. Coast. Shelf Sci. 14, 175 - 192.

Mulcahy, M.F., Twomey, E., Petersen, A and Maye, C.T. (1987) 'Pathobiology of estuarine fish and shellfish in relation to pollution' in D.H.S. Richardson (ed.) 'Biological Indicators of Pollution', Royal Irish Academy, Dublin, pp. 201 - 210.

Murray, C.N., Kautsky, H., Hoppenheit, M. and Domian, M. (1978) 'Actinide activities in water entering the North Sea', Nature, Lond. 276, 225-30.

NAS/NAE (1973) 'Water Quality Criteria 1972'. U.S. Environmental Protection Agency, Washington D.C.

Nedwell, D.B. and Abram, J.W. (1978) 'Bacterial sulphate reduction in relation to sulphur geochemistry in two contrasting areas of saltmarsh sediment'. Est. Coast. Mar. Sci. 6, 341 - 351.

Neinhuis, P.H. (1978) 'An ecosytem study in Lake Grevelingen, a former estuary in the SW Netherlands'. Keil. Meeresforsch. 4, 247 - 255.

Nelson-Smith, A. (1977) 'Biological consequences of oil spills' in J. Lenihan and W.W. Fletcher (eds.) 'The Marine Environment', Blackie, Glasgow, pp. 46 - 69.

NERC (1974) 'The Clyde estuary and Firth. An assessment of present knowledge compiled by members of the Clyde Study Group'. NERC Publications Series C, Number 11.

Nicholas, P.C. (1985) 'Predictive modelling in estuary quality management' in J.G. Wilson and W. Halcrow (eds.) 'Estuarine Management and Quality Assessment', Plenum Press, London, 203 -

218.

Nicholls, F.H. (1977) 'Infaunal biomass and production on a mudflat, San Fancisco Bay, California' in B.C. Coull (ed.) 'Ecology of marine benthos', University of South Carolina Press, Columbia.

Odum, E.P. (1980) 'The status of three ecosystem-level hypotheses regarding salt marsh estuaries: tidal subsidy, outwelling, and detritus-based food chains' in V.S. Kennedy (ed.) 'Estuarine Perspectives', Academic Press, New York, pp. 485 -495.

Odum, E.P. and Fanning, M.E. (1973) 'Comparisons of the productivity of *Spartina alterniflora* and *S. cynosuroides* in Georgia coastal marshes'. Bull. Georgia Acad. Sci. 31, 1 - 12.

Odum, E.P. and Smalley, A.E. (1959) 'Comparison of the population energy flow of a herbivorous and deposit- feeding invertebrate in a salt marsh ecosystem'. Proc. Nat. Acad. Sci. USA 45, 617 - 622.

Odum, H.T. and Copeland, B.J. (1974) 'A functional classification of the coastal systems of the United States' in H.T. Odum, B.J. Copeland and E.A. McMahan (eds.) 'Coastal Ecological Systems of the United States', The Conservation Foundation, Washington, D.C., 5 - 71.

O'Halloran, J., Myers, A.A. and Duggan, P.F. (1987) 'Lead poisoning in mute swans and fishing practice in Ireland' in D.H.S. Richardson (ed.) 'Biological Indicators of Pollution', Royal Irish Academy, Dublin, pp. 183 - 191.

O'Kane, J.P. (1980) 'Estuarine Water Quality Management', Pitman Advanced Publishing Program, Boston.

Oslo and Paris Commissions (1984) 'The First Decade: International Co-operation in Protecting our Marine Environment'. Oslo and Paris Commissions, London.

O'Sullivan, A.J. (1987) 'Recreation' in M. Bruton, F.J. Convery and A. Johnson (eds.), 'Managing Dublin Bay', REPC, Dublin, pp. 112 - 128.

Page, T. (1983) 'Uncertainty and policy formation' in E.P. Myers (ed.) 'Ocean disposal of municipal waste water: impacts on the coastal environment', Massachussetts Institute of Technology, Cambridge, Mass., pp. 613 - 57.

Paine, R.T. (1977) 'Controlled manipulations in the marine intertidal zone, and their contributions to ecological theory' in C.E. Goulden (ed.) 'The Changing Scenes in Natural Sciences 1776 -

1976', Acad. Nat. Sci. Spec. Pub. 12, Philadelphia, pp. 245 - 270.

Parry, W.K. and Adeney, W.E. (1901) 'The discharge of sewage into a tidal estuary'. Proc. Inst. Civ. Eng. 147, 70 - 154.

Parsons, T.R., Maita, Y. and Lalli, C.M. (1984) 'A Manual of the Chemical and Biological Methods for Seawater Analysis', Pergamon Press, Oxford.

Patriquin, D.G. (1972) 'The origin of nitrogen and phosphorus for growth of the marine angiosperm *Thalassia testudinum*'. Mar Biol. 15, 35 - 46.

Patriquin, D.G. (1978) 'Nitrogen fixation (acetylene reduction) associated with cord grass *Spartina alterniflora* Loisel'. Ecol. Bull. (Stockholm) 26, 20 - 27.

Pearson, T. H. (1982) 'The Loch Eil project: assessment and synthesis with a discussion of certain biological questions arising from a study of the organic pollution of sediments'. J. exp. mar. Biol. Ecol. 57, 93 - 124.

Pearson, T.H. and Rosenberg, R. (1978) 'Macrobenthic succession in relation to organic enrichment and pollution of the marine environment'. Oceanogr. Mar. Biol. Ann. Rev. 16, 229 - 311.

Phillips, D.J.H. (1980) 'Quantitative Aquatic Biological Indicators', Applied Science Publishers, Barking, Essex.

Pike, E.B. (1975) 'Aerobic bacteria' in C.R. Curds and H.A. Hawks (eds.) 'Ecological Aspects of Used Water Treatment', Pergamon Press, London, pp. 1 - 58.

Platt, H.M. and Lambshead, P.J.D. (1985) 'Neutral model analysis of patterns of marine benthic species diversity'. Mar. Ecol. Prog. Ser. 24, 75 - 81.

Pomeroy, R.D. (1959) 'The empirical approach for determining the required length of an ocean outfall' in E.A. Pearson (ed.) 'Proceedings of the 1st International Conference on Waste Disposal in the Marine Enviroment', University of California, Berkeley, Pergamon Press, New York, pp. 268 - 278.

Porter, E. (1973) 'Pollution in industrialised estuaries. Studies in relation to changes in population and industrial development'. Royal Commission on Environmental Pollution No. 4, HMSO, London.

Portmann, J.E. and Wood, P.C. (1985) 'The UK national estuarine classification scheme and its application' in J.G. Wilson and W. Halcrow (eds.) 'Estuarine Management and Quality

Assessment', Plenum Press, New York, pp. 173 - 186.

Preston, F.W. (1948) 'The commonness and rarity of species'. Ecology 29, 254 - 283.

Pritchard, D.W. (1955) 'Estuarine circulation patterns'. Am. Soc. Civil Eng., 81, 717/1 - 717/11.

Quetin, B. and de Rouville, M. (1986) 'Submarine sewer outfalls'. Mar. Pollut. Bull. 17, 133 - 183.

Radford, P.J. (1981) 'Modelling the impact of a tidal power scheme upon the Severn estuary ecosystem'. 1981 International Symposium on Energy and Ecology Modelling, Louisville, Kentucky, 1981.

Rankin, J.C. and Davenport, J. (1981) 'Animal Osmoregulation'. Blackie, Glasgow.

Ranwell, D.S. (1961) ' marshes in southern England. 1. The effects of sheep grazing at the upper limits of marsh in Bridgwater Bay'. J. Ecol. 49, 325 - 340.

Reimold, R.J. (1972) 'The movement of phosphorus through the salt marsh cord grass *Spartina alterniflora* Loisel'. Limnol. Oceanogr. 17, 606 - 611.

Remane, A. (1971) 'Ecology of brackish water' in A. Remane and C. Schlieper 'Die Binnengewasser', Band 25, Wiley Interscience, New York.

Riddle, A.M. (1985) 'Observations and mathematical model for the Wyre estuary' in J.G. Wilson and W. Halcrow (eds.) 'Estuarine Management and Quality Assessment', Plenum Press, London, pp. 27 - 36.

Riley J.P. and Chester, R. (1971) 'Introduction to Marine Chemistry'. Academic Press, New York.

Rowe, G.T. Clifford, C.H., Smith, K.L. and Hamilton, P.L. (1975) 'Benthic nutrient regeneration and its coupling to primary productivity in coastal waters'. Nature (Lond.) 255, 215 - 217.

Rygg, B. (1986) 'Heavy-metal pollution and log-normal distribution of individuals among species in benthic communities'. Mar. Pollut. Bull. 17, 31 - 36.

Saeijs, H.L.F. and Stortelder, P.B.M. (1982) 'Converting an estuary to Lake Grevelingen: environmental review of a coastal engineering project'. Environ. Management 6, 377 - 406.

Sand-Jensen, K. (1975) 'Biomass, net production and growth dynamics in an eelgrass (*Zostera marina*) population in Vellerup Vej, Denmark'. Ophelia,

14, 185 - 201.

Sasamura, Y. (1981) ' Petroleum in the Marine Environment', IMCO, London.

Sayers, D.R. (1986) 'Derivation and application of environmental quality objectives and standards to discharges to the Humber estuary (U.K.)'. Wat. Sci. Technol. 18, 277 - 285.

Schlieper, C. (1971) 'Physiology of brackish water' in A. Remane and C. Schlieper 'Die Binnengewasser', Band 25, Wiley Interscience, New York.

Shaw, K.M., Lambshead, P.J.D. and Platt, H.M. (1983) 'Detection of pollution-induced disturbance in marine benthic assemblages with special reference to nematodes'. Mar. Ecol. Prog. Ser. 11, 195 - 202.

Simkiss, K. and Schmidt, G.H. (1985) 'Cell mediated responses to metals - three novel approaches'. Mar. Environ. Res. 17, 188 - 191.

Soltanpour-Gargari, A. and Wellershaus, S. (1985) 'Eurytemora affinis - one year study of abundance and environmental factors'. Veroff. Inst. Meeresforsch. Bremerh. 20, 183 - 198.

Soulsby, P.G., Lowthion, D. and Houston, M.C.M. (1985a) 'The identification and evaluation of environmental quality in Southampton Water, UK, using limited manpower resources' in J.G. Wilson and W. Halcrow (eds.) 'Estuarine Management and Quality Assessment', Plenum Press, London, pp. 187 - 202.

Soulsby, P.G., Lowthion, D., Houston, M.C.M. and Montgomery, H.A.C. (1985b) 'The role of sewage effluent in the accumulation of macroalgal mats on intertidal mudflats in two basins in Southern England'. Neth. J. Sea Res. 19, 257 - 263.

Spedding, C.R.W. (1977) 'The Biology of Agricultural Systems', Academic Press, London.

Stebbing, A.R.D. (1976) 'The effects of low metal levels on a clonal hydroid'. J. mar. biol. Ass. U.K. 56, 977 - 994.

Stebbing, A.R.D. (1979) 'An experimental approach to the determinants of biological water quality'. Phil. Trans. R. Soc. Lond. B 286, 465 - 481.

Stebbing, A.R.D. (1981a) 'The effects of reduced salinity on colonial growth and membership in a hydroid'. J. exp. mar. Biol. Ecol. 55, 233 - 241.

Stebbing, A.R.D. (1981b) 'The kinetics of growth control in a colonial hydroid'. J. mar. biol. Ass. U.K. 61, 35 - 63.

Steele, J.H. (1974) 'The Structure of Marine

Ecosystems', Harvard University Press, Massachussetts.

Taylor, D. (1977) 'A summary of the data on the toxicity of various metals to aquatic life. 1. Mercury'. ICI, Brixham, Devon Report No. BL/A/1784.

Taylor, D. (1981a) 'A summary of the data on the toxicity of various metals to aquatic life. 9. Arsenic'. ICI, Brixham, Devon Report No. BL/A/2120.

Taylor, D. (1981b) 'A summary of the data on the toxicity of various metals to aquatic life. 5. Copper'. ICI, Brixham, Devon Report No. BL/A/1900.

Taylor, D. (1981c) 'A summary of the data on the toxicity of various metals to aquatic life. 10. Lead'. ICI, Brixham, Devon Report No. BL/A/2126.

Taylor, D. (1981d) 'A summary of the data on the toxicity of various metals to aquatic life. 13. Zinc'. ICI, Brixham, Devon Report No. BL/A/2143.

Thayer, G.W., Adams, S.M. and La Croix, M.W. (1975) 'Structural and functional aspects of a recently established *Zostera marina* community' in L.E. Cronin (ed.) 'Estuarine Research' Vol. 1, Academic Press, New York, pp. 518 -540.

Theede, H. (1980) 'Physiological responses of estuarine animals to cadmium pollution'. Helgolander Meeresunters. 33, 26 - 35.

Tomlinson, D.L., Wilson, J.G., Harris, C.R. and Jeffrey, D.W. (1980) 'Problems in the assessment of heavy metal levels in estuaries and the formation of a pollution index'. Helgolander. Meeresunters. 33, 566 - 575.

Tripp, M.R., Fries, C.R., Craven, M.A. and Grier, C.E. (1984) 'Histopathology of *Mercenaria mercenaria* as an indicator of pollution stress'. Mar. Environ. Res. 14, 521 - 524.

Valiela, I. and Teal, J.M. (1974) 'Nutrient limitation in salt marsh vegetation' in R.J. Reimold and W.H. Queen (eds.) 'Ecology of Halophytes', Academic Press, New York, pp. 547 - 563.

Valiela, I. and Teal, J.M. (1979) 'Inputs, ouputs and inter- conversions of nitrogen in a salt marsh ecosystem' in R.L. Jefferies and A.J. Davy (eds.) 'Ecological Processes in Coastal Environments', Blackwell, Oxford, pp. 399 - 414.

van de Meent, D, den Hollander, H.A., Pool, W.G., Vredenbregt, M.J., van Oers, H.A.M., de Greef, E. and Luitjen, J.A. (1986) 'Organic micropollutants in Dutch coastal waters'. Wat.

Sci. Technol. 18, 73 - 82.

van Impe, J. (1985) 'Estuarine pollution as a probable cause of increase of estuarine birds'. Mar. Pollut. Bull. 16, 271 - 276.

van Pagee, J.A., Gerritsen, H. and de Ruijter, W.P.M. (1986) 'Transport and water quality in the southern North Sea in relation to coastal pollution research and control'. Wat. Sci. Tech. 18, 245 - 256.

Waite, T.D. (1984) 'Principles of Water Quality', Academic Press, London.

Warwick, R.M., Joint, I.R. and Radford, P.J. (1979) 'Secondary production of the benthos in an estuarine environment' in R.L. Jefferies and A.J. Davy (eds.) 'Ecological Processes in Coastal Ecosystems', Blackwell, Oxford, pp. 429 - 450.

Warwick, R.M. and Price, R. (1975) 'Macrofauna production on an estuarine mudflat'. J. mar. Biol. Ass. U.K. 55, 1 - 18.

Webb, K.L., DuPaul, W., Weibe, W.J., Sottile, W., and Johannes, R.E. (1975) 'Aspects of the nitrogen cycle on a coral reef'. Limnol. Oceanogr. 20, 198 - 210.

Wellershaus, S. (1981) 'Turbidity maximum and mud shoaling in the Weser estuary'. Arch. Hydrobiol. 92, 161 - 198.

Welsh, B.L. (1975) 'The role of the grass shrimp *Palaemonets pugio* in a tidal marsh ecosystem'. Ecology 56, 513 - 30.

West, A.B., Partridge, J.K. and Lovitt, A. (1979) 'The cockle *Cerastoderma edule* (L) on the South Bull: population parameters and fisheries potential'. Ir. Fish. Invest. Ser. B, 20, 1 - 18.

Wheeler, W.N. (1978) 'Ecophysiological studies on the giant kelp, *Macrocystis*'. Ph.D. Thesis, University of California, California.

Whitney, D.E., Woodwell, G.M. and Howarth, R.W. (1975) 'Nitrogen fixation in Flax Pond: A Long Island salt marsh'. Limnol. Oceanogr. 20, 640 - 643.

WHO/UNEP (1982) 'Waste discharge into the marine environment. Principles and guidelines for the Mediterranean action plan'. Pergamon Press, Oxford.

Wilson, J.G. (1976) 'Upper temperature tolerances of *Tellina tenuis* and *Tellina fabula*' Mar. Biol. 45, 123 - 128.

Wilson, J.G. (1983) 'The uptake and accumulation of Ni by *Cerastoderma edule* and its effects on

mortality, body condition and respiration rate'.
Mar. Environ. Res. 8, 129 - 148.
Wilson, J.G., Allott, N., Bailey, F. and Gray, N.F.
(1986) 'A survey of the pollution status of the
Liffey estuary'. Ir. J. Environ. Sci. 3, 15 -
20.
Wilson, J.G., Ducrotoy, J.-P., Desprez, M. and
Elkaim, B. (1987) 'Application d'indices de
qualite ecologique des estuaires en Manche
centrale et orientale'. Vie Milieu 37, 1 - 11.
Wilson, J.G. and Earley, J.J. (1985). 'Pesticide and
PCB levels in the eggs of shag (*Phalacrocorax
aristotelis*) and cormorant (*P. carbo*) from
Ireland'. Envir. Pollut. Ser. B 12, 15 - 26.
Wilson, J.G. and Jeffrey, D.W. (1987) 'Europe-wide
indices for monitoring estuarine quality' in
D.H.S. Richardson (ed.) 'Biological Indicators
of Pollution', Royal Irish Academy, Dublin, pp.
225 - 242.
Wilson J.G. and McMahon, R.F. (1981) 'Effects of
high environmental copper concentration on the
oxygen consumption, condition and shell
morphology of natural populations of *Mytilus
edulis* L. and *Littorina rudis* Maton'. Comp.
Biochem. Physiol. 70C, 139 - 147.
Wolff, W.J. (1977) 'A benthic food budget for the
Grevelingen estuary, The Netherlands, and a
consideration of the mechanisms causing high
benthic secondary production in estuaries' in
B.C. Coull (ed.) 'Ecology of Marine Benthos',
University of South Carolina Press, Columbia.
Wolff, W.J. and de Wolff, L. (1977) 'Biomass and
production of zoobenthos in the Grevelingen
estuary, The Netherlands'. Estuar. Coast. Mar.
Sci. 5, 1 - 24.
Woodwell, G.M., Houghton, R.A., Hall, C.A.S.,
Whitney, D.E., Moll, R.A. and Juers, D.W. (1979)
'The Flax Pond ecosystem study: exchanges of
phosphorus between a salt marsh and the coastal
waters of Long Island Sound' in R.L. Jefferies
and A.J. Davy (eds.) 'Ecological Processes in
Coastal Environments', Blackwell, Oxford, pp.
491 - 511.
WRC, (1977) 'Cost information for water supply and
sewage disposal'. Water Research Centre,
Medmenham, TR 61.

Index